An Alternative to Oil:

BURNING COAL WITH GAS

*The economic and environmental benefits
of burning coal and natural gas mixtures
in boilers originally designed for oil*

An Alternative to Oil:

BURNING COAL WITH GAS

*The economic and environmental benefits
of burning coal and natural gas mixtures
in boilers originally designed for oil*

A.E.S. Green, *editor*

With contributions by

J. R. Jones, Jr.
Assistant Chairman, Department of Reference and Bibliography

M. J. Ellerbrock
Assistant Professor, Department of Food and Resource Economics, IFAS

J. M. Schwartz
Electrical Engineer, Executive Officer, ICAAS

S. J. Kuntz
Graduate, Department of Economics, College of Business

B. Zeiler
Graduate, Department of Mechanical Engineering

A University of Florida Book

UNIVERSITY PRESSES OF FLORIDA
famu / fau / fiu / fsu / ucf / uf / unf / usf / uwf
GAINESVILLE

University Presses of Florida is the central agency for scholarly publishing of the State of Florida's university system. Its offices are located at 15 NW 15th Street, Gainesville, FL 32603.

Works published by University Presses of Florida are evaluated and selected for publication, after being reviewed by referees both within and outside of the state's university system, by a faculty editorial committee of any one of Florida's nine public universities: Florida A&M University (Tallahassee), Florida Atlantic University (Boca Raton), Florida International University (Miami), Florida State University (Tallahassee), University of Central Florida (Orlando), University of Florida (Gainesville), University of North Florida (Jacksonville), University of South Florida (Tampa), University of West Florida (Pensacola).

An Alternative to oil.

Includes bibliographical references and indexes.
1. Boilers--Fuel. 2. Coal. 3. Gas, Natural.
I. Green, Alex Edward Samuel, 1919-
II. Jones, J. R. (Jesse R.)
TJ263.5.A37   621.1'82   82-1894
ISBN 0-8130-0724-0   AACR2

CONTENTS

PREFACE

This monograph is an interdisciplinary assessment of the economic
and environmental benefits of burning coal and natural gas mixtures in
boilers and furnaces originally designed for oil. It is written as a
companion to our earlier interdisciplinary monograph, *Coal Burning Issues,* to include what now appears to be a new way of displacing oil
and burning coal cleanly. The earlier study, as did many other energy
studies during the 1974–80 period, implicitly assumed that United
States resources of oil *and* gas were both depleting rapidly. Recent analyses, however, suggest that the U.S. has greater reserves of natural
gas than estimated in 1978. If gas production can hold up, at least for the
next two decades, then the use of our abundant supply of coal together with our moderate supply of gas will provide a faster and environmentally better way of reducing our nation's excessive dependence
upon foreign oil. Thus simultaneous gas-coal burning might provide the
opportunity to overcome the economic problems associated with our
large negative balance of trade and the attendant degradation of our quality
of life.

ACKNOWLEDGMENTS

This work was fostered by a grant from the American Gas Association to carry out a detailed analysis of the use of natural gas and coal in
Florida's electric utility system (see Chapter 1, Sections 1.2 and 1.6,
and Chapters 5, 6, 7, and 8 of this book) and an additional subsidy
supporting the publication of this broader work. The authors would like to
thank the Policy Evaluation and Analysis Group of the American
Gas Association, in particular Dr. Benjamin Schlesinger, Mr. Nelson Hay,
and Mr. Paul Wilkinson, not only for their financial support but also
for their enthusiastic encouragement and help.

Collateral to this policy study a development program and a scientific
laboratory program, initiated early in 1981, have had a strong influence on this monograph. These two programs have in large part been
supported by University of Florida funds. Special thanks are due to
J.A. Nattress, W.H. Chen, M.J. Ohanian, and F.G. Stehli for these
arrangements. Additional support in the form of funding, services, facilities, or materials, or privileged technical information has been given
by the Department of Physics' instrument shop, Continental Resources
Company (Winter Park, Florida), Middle Ultraviolet Associates
(Gainesville, Florida), Dragon Fire and Clay Company (Micanopy, Florida), and the General Shale Products Corporation (Johnson City,
Tennessee). Donations of coal and other material used in our experiments
have been made by the Florida Power and Light Corporation (Miami, Florida), Gainesville Regional Utilities (Gainesville, Florida), Virginia Fuel Company (Chattanooga, Tennessee), North River Energy

Company (Berry, Alabama), Drummond Coal Company (Jasper, Alabama), and Estech General Chemicals Corporation (Bartow, Florida).

The editor would like to thank the Energy Committee of the University of Florida Library, Dr. Douglas Pewitt and Mr. Alfred C. Metz of the U.S. Department of Energy, Dr. Robert E. Hall and W. Steven Lanier of the Environmental Protection Agency, Professor L.D. Smoot of Brigham Young University, Mr. Fred Hancock of Gainesville Regional Utilities, Dr. Michael Von Seebeck of the Polysius Corporation, M.R. Milner of the Trimex Corporation, my colleagues Professors Richard T. Schneider and Charles L. Proctor II, and my research associates Dr. Deepak B. Vaidya, Dr. Krishna Pamidimukkala, and Mr. John J. Horvath, who facilitated our acquisition of books, documents, and other materials that were vital to our studies.

The draft report of our summer study, dated 26 August 1981, was circulated rather widely for comments and criticism. We wish to thank the men and organizations who responded and hope that this monograph favorably reflects considerations of their comments and suggestions.

The editor of this monograph would also like to thank his coauthors for permitting him to imbed the careful collective quantitative study made in the summer of 1981 (chapters 5–8) and a report, "Gas-Coal Burning Options," into a broader work "fleshed out" by the editor. Lest the coauthors be blamed for any of the shortcomings of these additional chapters, however, it is appropriate to delimit their contributions: Chapter 2 is largely the work of Dr. J. Raymond Jones, Jr.; Chapter 5 is mainly the work of Bernhard Zeiler; Chapter 6 is mainly the work of Jerome M. Schwartz; Chapter 7 is exclusively the work of Dr. Michael J. Ellerbrock; Chapter 8 is mainly the work of Steven J. Kuntz.

Thanks are also due to Michael T. Crimmins for his valuable contributions to the organization of the summer study and to Jerome Schwartz and Bernhard Zeiler for their contributions to the organization of the additional chapters. In addition we thank Linda Green for the art work, Roxanne Wilkerson and Olivia Berger for manuscript typing, Hazel Feagle and Claire Myers of the Word Processing Center of the College of Engineering, University of Florida, for the production of the final copy, and Jeffery M. Samuels for computer support. The Board of Managers of the University of Florida Press and the staff of University Presses of Florida were unusually cooperative in our attempt to see this work published in the shortest possible time.

I am personally thankful to Professor Robert B. Gaither, chairman of the Department of Mechanical Engineering, for allowing a wandering physicist to teach a graduate course in combustion in his department. The opportunity to teach affords a great opportunity to learn, and it is hoped that this monograph reflects such experience. I would like also to thank the administrators of the University of Florida for their continued support of ICAAS even when its direction departed from its traditional topical area.

Finally I would like to thank my son Bruce, who, as a production potter, has been "playing with fire" for over a dozen years. His experience, judgment, and intuitive approach to our experimental firings strongly influenced my own decision to undertake studies of the simultaneous combustion of natural gas and coal.

Alex E.S. Green / December 1981

CHAPTER 1

BACKGROUND, OVERVIEW AND SUMMARY

1.1 Background

Recently the Interdisciplinary Center for Aeronomy and (other) Atmospheric Sciences (ICAAS) at the University of Florida completed two broad interdisciplinary studies relative to coal burning. The first, "Coal Burning Issues" (ICAAS 1980a, to be referred to as CBI), was a monograph written in a national context reporting the results of the scoping phase of an interdisciplinary assessment of the impact of the increased use of coal. The second report, entitled "The Impact of Increased Coal Use on Florida (IICUF)" (ICAAS 1980b), was a study to determine major economic, social, environmental and technological issues to be considered in reference to increased coal burning in Florida and a projection of options available to the state attendant to each issue. The major conclusion of both studies was that the nation and particularly the State of Florida must quickly reduce their large reliance on foreign oil and that conservation measures and increased reliance on the abundant national supply of coal were the major alternatives available to the public in the next few decades.

More recently, increased estimates of the economically recoverable natural gas reserves in the United States prompted the study, "Gas-Coal Burning Options" (Green and Jones 1981), which concluded that the State of Florida, and with it the United States, should broaden the options in its transitional energy plan to include the development and utilization of flexible combinations of gas-coal fuels for industrial and utility boilers. The coherence of this last study with economic analyses favoring select gas use with coal (American Gas Association 1981, Schlesinger 1980), led to a grant by the AGA to ICAAS to study economic and related technical aspects of converting Florida's electric utility oil boilers to gas-coal burning.

1

1.2 Abstract of Report to American Gas Association (26 August 1981) on the
    Conversion of Utility Oil Boilers in Florida to Gas and Coal Burning

"Natural gas production levels in the United States now appear sustainable to the year 2000 at 20 trillion cubic feet (TCF) or greater. Thus natural gas need not be phased out of utility and industrial applications as originally specified by the 1978 Power Plant and Industrial Fuel Use Act. This study evaluates the use of gas with coal in Florida to replace residual oil in utility steam boilers originally designed for oil burning. The conversion costs (in $/kilowatt) are estimated for various oil-backout options: (a) coal-oil mixtures, (b) dual coal or oil use, (c) new coal-fired boiler, (d) gas-coal, (e) water-coal, and (f) water-gas-coal. Simplified economic comparisons based upon the annual costs of a 200-megawatt unit favor the gas-coal option. A more refined and detailed economic analysis of the conversion to gas-coal of 12,678 megawatts of suitable oil boiler capacity in Florida is carried out, which allows for fuel and general inflation rates and economies of scale. The results indicate a savings of 2.5 billion dollars per year in annualized electric generation cost, in 1980 dollars, over the 1980-1995 period and a reduction of about 50 percent in the annualized power production cost.

The most important factor that influences our economic projections is the price difference between oil and coal energy. Gas energy, which plays only a 30 percent role, is not very influential upon costs. The air quality impact of conversion from oil to gas-coal is examined. It is shown to be possible to maintain or improve upon current emission standards. Oil use in Florida's ten-year electrical generation expansion plan would be substantially reduced if gas-coal conversions were implemented."

The special nature of Florida's problem is illustrated by Figure 1.1, which gives the 1980 United States Energy Consumption and Figure 1.2, which projects the 1980 Florida Energy Consumption. Whether gas and coal can displace the much larger proportion of oil used in electrical generation in Florida (~50%) as compared to the USA as a whole (~15%) was the underlying question of our Florida study.

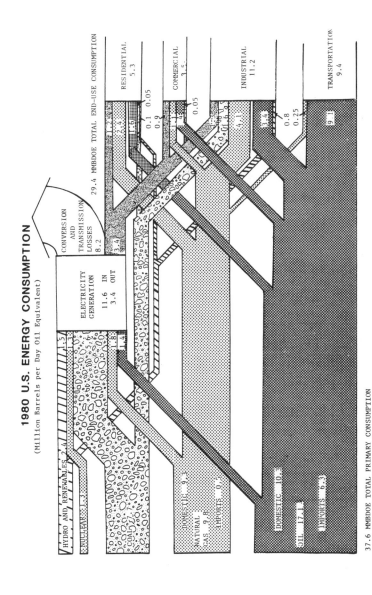

Fig. 1.1 United States Energy Supply and Disposition for 1980 (from Department of Energy 1980 Annual Report to Congress DOE/EIA-0173 (80)).

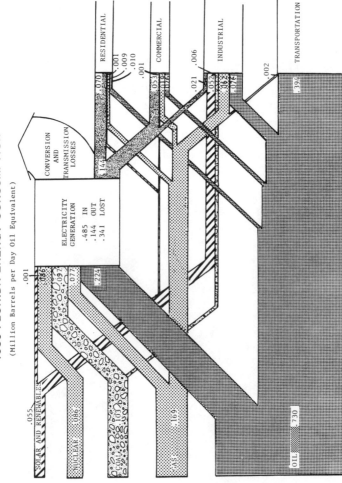

Figure 1.2

Fig. 1.2 Florida's Energy Supply and Disposition for 1980 (data from Forecasts of Energy Consumption in Florida 1980-2000, Sept. 1981, State of Florida Governor's Energy Office).

1.3 State of the Gas-Coal Burning Literature

About the same time as the initiation of the Green-Jones (1981) study, an experimental study of natural gas-pulverized coal burning was started using a burner in the 0.5 million BTU/hr range. This field program was influenced by two years of exploratory thinking and discussion (Green and Green 1980, 1981) on the use of coal to displace propane burning by a production pottery kiln. Shortly thereafter a laboratory study of gas-coal flames was initiated using a burner in the 10,000 BTU/hr range (Horvath, Vaidya and Green 1981). The primary intent here was to carry out gas-coal burning under carefully controlled conditions and to use the observations to help develop a theoretical model of this mixed combustion process. Most of the observations were carried out with the use of middle ultraviolet spectroscopy since refined instruments were available to utilize this specialty of the editor (Green 1966, 1981). The radiations emitted by various zones of gas-coal flames are being used to obtain information about the chemical, radiative and flow processes taking place in heterogeneous gas-coal flames.

These experimental and theoretical efforts, collateral to the AGA-ICAAS economic study. led to a rapid familiarization with the litera- tures of pulverized coal combustion and natural gas combustion, two vast technical and scientific literatures. This familiarization revealed a paucity of flame studies of natural gas-pulverized coal mixtures, apparently accounted for by the absence of applications for this fuel combination. While combustion studies have increased substantially in the United States in recent years, most of the new studies have been directed at answering the question of whether we can use our abundant supply of coal to develop alternative fuels to substitute for oil and natural gas? Attempts to answer this question largely derive from the 1978 Power Plant and Industrial Fuel Use Act (PIFUA). which reflected the national concern that the oil and natural gas reserves of the USA were depleting rapidly. The possibility that domestic natural gas supplies are not at great risk for the next 20 years or so has been raised only very recently. If one grants this possibility, then different questions arise: How can we use our abundant reserves of coal and our modest supplies of natural gas to displace oil so as to minimize our imports, a major source of our current

economic problems? While addressing this economic question, how can we
use natural gas to mitigate the environmental consequences of increased
coal burning? This monograph addresses these questions.

1.4 Coal and Natural Gas Supplies

It is generally recognized that the United States is well endowed
with coal reserves and that it has coal production capabilities
considerably in excess of current coal production levels. The gas supply
situation is, however, the subject of some controversy with substantial
differences appearing in estimates by authoritative groups. Figure 1.3
shows the projections of the Tennessee Gas Transmission Co. (TGT) 1981, a
more conservative view of our production capabilities through the year
2000. The various shaded areas from the bottom upward designate production from (A) proven reserves established by 1980 in the "Lower 48"
(B) future reserve additions in the Lower 48 (C) offshore Atlantic reserves (D) Alaskan reserves excluding liquefied natural gas (E) conventional pipeline imports from Canada and Mexico.

The curve immediately above the shaded area represent TGT's total
production estimate. This includes all nonconventional sources of natural
gas (R) such as liquefied natural gas from other foreign countries,
synthetic natural gas from coal, and the sum of other new technologies or
sources.

The upper lines with arrows on Fig. 1.3 give the 1981 estimates of
the American Gas Association under the assumption of a scenario which
emphasizes North American supply sources and considers a broader range of
nonconventional sources. The higher points of the range imply two major
assumptions. "First, that there is a vigorous national R&D effort to
improve the technologies related to gas production and, second, that
federal policy and regulatory decisions support, consistent with the
scenario description, increasing gas supplies. The low estimate points
are based on an indifferent national R&D program with dilatory federal and
regulatory policy decisions." (Wilkinson 1981). Both AGA and TGT estimates indicate that gas supplies to the year 2000 will be available at the
20 TCF/yr level, a level also accepted in the most recent National Energy
Plan.

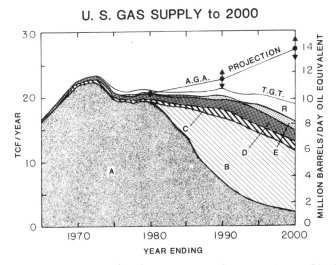

Fig. 1.3 U.S. Gas Supply Projections to 2000 (Adapted from p.18, Energy 1901-2000, the Tennessee Gas Transmission Company, Houston, Texas, August 1981). The upper straight line segments represent the American Gas Association estimates under a scenario which emphasizes North American supply sources. The arrows indicate the range of estimates.

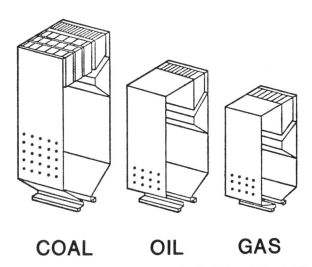

Fig. 1.4 Relative Sizes and Features of Coal, Oil and Gas Boilers (from Erhlich, S., S. Drenker and R. Manfred, Coal Use in Boilers Designed for Oil and Gas Firing, American Power Conference, Chicago, Illinois, April 21-23, 1980.

This 20 TCF level of gas supply has considerable significance from the viewpoint of the gas-coal-oil displacement thesis. Table 1.1 summarizes the gas and coal supplies per year needed to displace various levels of oil use. Our current annual coal production is at a rate of about 700 million tons per year and our production capability is about 800 million tons. From the entries in the table it should be obvious that even a displacement of 3-5 million barrels of oil per day could readily be accommodated with only minor or modest increases in our domestic gas and coal production capabilities. With some redistribution of our current gas uses it should be possible to accommodate even greater levels of oil displacement. Accordingly the gas-coal-oil displacement approach can be accommodated from a supply standpoint.

1.5 A Summary of the Physical Basis of Gas-Coal Replacement of Oil

Chapter 4, The Physical Basis of Gas-Coal Replacement of Oil, discusses various physical reasons in support of the concept that gas and coal can displace oil in utility boilers. These reasons rest largely upon the intermediate nature of various properties of residual oil as compared to gas and coal. Figure 1.4 (adapted from Ehrlich, Drenker and Manfred 1980) illustrates this intermediate character of oil with a display of the relative sizes and boiler design differences for coal (bituminous), oil (residual) and gas (natural) boilers which produce equal steam power output. Other examples of such properties are summarized in Table 1.2. Because of the intermediate characteristics, it is reasonable to expect in all matters with the exception of ash problems that a gas-coal mixture fed into the burning zone of an oil boiler would burn somewhat like oil. Ash problems such as slagging, fouling, increased corrosion, tube erosion, plugging of air passes and increased particulate emission require more complex measures (Babcock and Wilcox 1978, Combustion Engineering 1981). The particulate emissions problem can be handled by the use of highly efficient precipitators or bag houses. Means of dealing with the other expected ash problems are just being developed as various approaches to

Table 1.1 Gas and Coal needs for various oil displacements
(using a 30/70 by energy gas/coal ratio)

μ	Oil Displaced Quads per yr	Gas Required TCF per yr	% Nat. Prod. (20 TCF)	Coal Required Million Tons per yr	% Nat. Prod. (800 10^6 T/yr)
1	2.2	0.66	3.3	64	8.0
2	4.4	1.31	6.6	128	16.0
3	6.6	1.97	9.8	192	24.0
5	11.0	3.28	16.4	319	39.9
7	15.3	4.60	23.0	447	55.9

μ = million barrels of oil per day $\approx$ 2.2 quad per yr
TCF = trillion cubic feet $\approx$ 1 quad
Quad = 10^{15} BTU $\approx$ 41.7 million tons of coal

Table 1.2   Characteristics of Gas, Oil and Coal

	Gas	Oil	Coal
Relative power density	1.6	1.0	0.73
Calorific value (10^3 BTU/lbs)	24	18	12
Air/fuel ratio by weight	20:1	17:1	11:1
Hydrogen/Carbon by weight by atom	1:3 4	1:9 1.6	1:20 0.8
SO_2 emission (lbs/10^6 BTU)	0	0.5 - 3	1 - 6
$CO_2:H_2O$ emissions	0.5:1	1.2:1	2.5:1
Burning time	fast	intermediate	slow
Radiation	low	middle	high
State of matter	gas	liquid	solid
Ash levels (percent)	0	0.2 - 0.5	1 - 10

oil boiler retrofitting are now being examined, tested or carried out.
These approaches include (1) conversion to coal oil mixtures such as
recently undertaken at the Sanford Station in Florida (Cook 1980),
(2) conversion to coal use with the maintenance of oil firing capability
as recently carried out at the Kwinana power station in Perth, Australia
(Kirkwood et al. 1978), (3) the installation of a new coal boiler to be
used with the remainder of the electric plant (Philipp 1979, 1980,
Ehrlich, Drenker and Manfred 1980) and (4) the use of coal-water slurries
(ibid., Glenn 1979). In addition various types of clean fuels derived
from coal using hydroliquefaction processes or chemical coal cleaning or
coal gasification have been proposed, but for the most part these will not
be available until the 1990s.

No prior literature appears to exist on the use of gas and coal for
oil boiler retrofitting, and it is hoped that this monograph will help
fill this void. The measures being used for coping with ash handling
problems for the first four options described above would also be applicable. In addition there is the possibility of using a first stage gas-coal
combustion chamber, which separates out a large portion of the ash before
the hot fuel-rich combustion gases including coal volatiles, CO, and other
gases are injected into the boiler for afterburning with natural gas
enhancement and secondary and tertiary air.

1.6 Economic Analyses

Chapter 5, Conversion Costs of Oil Back-Out Options, first classifies
all of Florida's utility oil boilers as to their suitability for coal
conversion. These classes are

A: Oil boilers located where existing or projected coal units are
     available.

B: Oil boilers in service approximately 25 years or less with large
     enough land area or port facilities to accommodate coal and ash
     handling.

C: Oil boilers in service approximately 25 years or less on small
     sites with no port facilities.

D: All remaining steam turbine oil boilers that have been in service
     more than 25 years.

E: All gas turbines which use distillate oil.

Classes D and E were not considered for conversion.

A methodology and simple economic analysis developed on retrofitting oil boilers to (a) coal-oil mixtures, (b) dual oil and coal capability with no flue gas desulfurization (FGD) and (c) installation of a new coal boiler was already available (Philipp 1980a,b). This methodology is extended to consider gas-coal, water-coal and gas-coal-water conversions. The total conversion cost per kilowatt for the various back-out options that should be of general applicability are summarized in Table 1.3. Breakdowns of these costs are given in Chapter 5. The third column of Table 1.3 gives annual savings of a 200 megawatt unit following such a conversion, and the fourth column gives the payback period for the capital costs of conversion. It is seen that these results favor the gas-coal option, to a large extent with respect to the coal-oil mixture option, and to varying lesser extents with respect to the other options.

Chapter 6, Quantitative Inputs for Economic Analyses, presents the inputs used in the final economic analyses and projects and quantifies the inflation rate and oil, gas, and coal growth rates. Figure 1.5 summarizes the essential features of the time dependences incorporated into the study. The large differences anticipated between oil and coal prices account for the major savings attendant to the displacement of oil by gas and coal. The price of gas which only supplies about 30 percent of the energy does not greatly influence the results.

Chapter 7, A Detailed Economic Analysis of Conversion Alternatives, discusses (1) the costs of capital, (2) the discount rate, (3) shutdown periods, (4) future fuel prices, and (5) payback period. The discussion is of general applicability and sets down the basis for our specific Florida calculations.

Chapter 8, Assessment of the Gas-Coal Conversion Option for Florida, presents the specific results of our analyses of the Florida utility system. The principal results are summarized in Table 1.4. A savings of 2.5 billion dollars per year is obtained by the conversion of 12,678 megawatts of oil boiler capacity. The reduction in annual oil use by Florida utilities would be about 0.2 million barrels of oil per day.

Table 1.3   Total Conversion Costs for Various Options

	Percent Coal Energy	Total Conversion Cost $/KW	Total Annual Cost 10^6/yr	Differential Annual Cost 10^6/yr	Payback Period
Oil	0	--	46.0	Base	--
COM	30	84	43.9	-2.1	3.3
Dual NFGD Coal	80	144	29.5	-16.5	1.5
Oil FGD	80	216	34.4	-11.6	2.4
NCFBWFGD[+]	100	376	36.0	-10.0	3.2
Gas/Coal	70	127	25.8	-20.2	1.2
Coal/Water	100	132	31.3*	-14.7	1.5
Gas-Coal/Water	70	132	29.4*	-16.6	1.4

[+] denotes new coal fired boiler with flue gas desulfurization.

* Includes extra energy cost of water evaporation and fine coal grind but not cost of peaking unit to replace lost capacity.

Table 1.4   Summary of Economic Gain for
Gas-Coal Conversion of Florida Oil Boilers

	Oil Base Case	Gas-Coal
Capacity (MW)	12,678	12,678
Conversion Costs (million dollars)		1429
Annualized Capital Cost (15 yrs @ 11%)		199
Annualized (extra 0 and M costs)		29
Total Annualized Costs	4,824	2,294
Savings from base case		2,530
Production Cost/KWH	7.2¢	3.4¢

Chapters 5, 6, 7, and 8 constitute the core of the detailed economic analyses on the conversion of utility oil boilers in Florida to gas and coal burning carried out under the AGA-ICAAS study. The fact that the pay-off of the gas-coal-oil back-out approach was much bigger than had been expected stimulated us to broaden our work to serve a national audience. To a large extent the other chapters of this monograph were developed for this purpose.

1.7 Air Quality Impacts

Chapter 9: Air Quality Impacts, considers the environmental impacts of oil-to-coal conversions that can arise because of the vastly greater ash content of coal and the somewhat greater sulfur content of many coals. Table 1.5 gives the characteristics of several of the major fuel groups, along with their energy release, ash and SO_2 emissions. The installation of a flue gas desulfurization (FGD) unit is extremely costly, and there usually is little space to accommodate it in a preexisting plant. Fortunately, it is possible to achieve acceptable levels of SO_2 by four simple and relatively inexpensive measures. These include

(1) the choice of a high quality or moderate quality coal

(2) the use of physical coal cleaning (PC)

(3) the burning of coal with natural gas (NG)

(4) the use of burner or boiler scrubbing (BS).

Figure 1.6 is a nomogram for calculating the emission levels for any fuel when the percentage sulfur by weight and the calorific values (in 10^3 BTU/lb) are specified. The lines illustrate the major fuel groups listed in Table 1.5. Note that high sulfur residual oil leads to much greater emissions than two of the coal groups (CAR and W) and also gives emissions comparable to the third group (NA). Thus, while the reductions factors achievable by measures 2, 3 and 4 are moderate, the combined reduction factor can bring a unit without FGD into the range achievable with Best Available Control Technology (BACT), which is usually accepted as 1.2 lbs SO_2/million BTU. This is illustrated in Table 1.6, which shows the effects of the reduction factors R(PC), R(NG), and R(BS) upon the final emission for various coal grades as defined by the normal emissions.

Table 1.5   Major Fuel Groups Uncontrolled SO_2 and TSP Emission Rates

FUEL	BTU CONTENT BTU/UNIT	%S	EMISSION	ASh %	EMISSION
COAL					
Northern Appalachian (NA)	12,000/LB	2.5	4.17 σ	14.0	7.58 π
Central Appalachian & Rockies (CAR)	12,000/LB	0.7	1.17 σ	12.0	6.50 π
Midwestern (MW)	11,000/LB	3.3	6.0 σ	11.0	6.50 π
Western (W)	8,500/LB	0.5	1.18 σ	9.0	6.88 π
RESIDUAL OIL					
High Sulfur (HS)	150,560/GAL	3.0	3.14 σ	-	0.053 π
Low Sulfur (LS)	146,430/GAL	0.3	0.31 σ	-	0.055 π
NATURAL GAS (NG)	1,027/FT^3	-	0.0006 σ	-	<0.015 π

σ denotes lbs $SO_2/10^6$BTU, π denotes lbs particulates/10^6BTU.
Adapted from ICF, Inc. Boilers System Cost Estimates, Table II-3, 1978

Table 1.6   Emissions (in lbs $SO_2/10^6$ BTU) for Various Coals, Physical Cleanings, Natural Gas and Burner Scrubbing.

R(PC)	R(NG)	R(BS)	A(1.2σ)	B(2.4σ)	C(4.2σ)	D(6.0σ)
1.0	1.0	1.0	1.2	2.4	4.2	6.0
		.6	0.72	1.44	2.52	3.6
	.75	1.0	0.9	1.8	3.15	4.5
		.6	0.54	1.08	1.89	2.7
	.60	1.0	0.72	1.44	2.52	3.6
		.6	0.43	0.86	1.51	2.16
	.50	1.0	0.6	1.2	2.1	3.0
		.6	0.36	0.72	1.26	1.8
.6	1.0	1.0	0.72	1.44	2.52	3.6
		.6	0.43	0.86	1.51	2.16
	.75	1.0	0.54	1.08	1.89	2.7
		.6	0.32	0.65	1.13	1.62
	.60	1.0	0.43	0.86	1.51	2.6
		.6	0.26	0.52	0.91	1.18
	.50	1.0	0.36	0.72	1.26	1.8
		.6	0.22	0.43	0.76	1.08

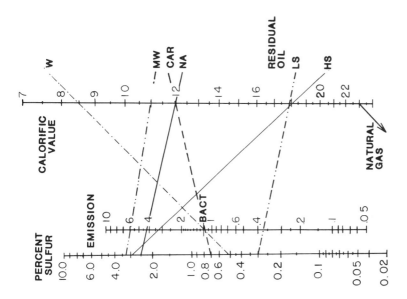

Fig. 1.6 Nomogram for Calculating Emission Levels (adapted from EPA-600/9-77-017). Emission levels are in $\sigma = ^{31}$ lb $SO_2/10^6$ BTU. Calorific values are in 10^3 BTU/lb.

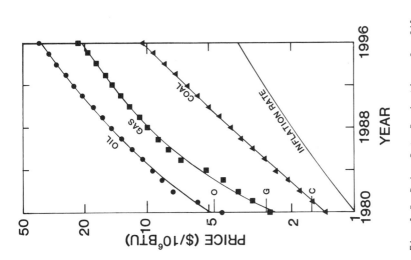

Fig. 1.5 Price Rate Projections for Oil, Gas and Coal and Inflation from AGA Tera Model The symbols O, G, and C indicate 1980 prices used in Chapter 8.

The technology for coping with particulates using precipitators or bag houses and ash handling boiler modifications is quite advanced and the costs of refitting, while large, are tolerable. In addition, it should be possible to minimize ash problems by using a first stage gas-coal combustor designed to remove ash. This approach will be discussed further in Chapter 11 on research and development needs for gas-coal displacement of oil.

1.8 National Assessment

Chapter 10, A National Assessment of Oil Displacement using Gas and Coal Burning, may be placed in perspective by Figure 1.1 and Table 1.7, which gives the trend of U.S. energy supply and disposition. The national capacity of utility oil boilers corresponding to the categories in our Florida study constitutes 63,279 megawatts. Proportioning this to the 12,678 Florida oil capacity and the 2.5 billion dollar per year Florida consumer saving, we arrive at a 12.5 billion dollar per year national savings. The 0.2 μ (million barrels of oil per day) saved by Florida extrapolates to 1.0 μ. The value of this savings to the U.S. as a country is, of course, greater than the consumer savings since we are keeping our dollars for gas and coal at home.

The industrial oil use of 3.4 μ constitutes an even more tempting alternative fuels target. Fortunately, many of the considerations applicable to smaller utility boilers are directly applicable to large industrial boilers. Furthermore, a number of national studies addressing the question of "Replacing Oil and Natural Gas by Coal in Industry" can be utilized to help provide approximate answers to the question of "Replacing Oil by Coal and Natural Gas in Industry".

On the basis of our Florida utilities study, it is not unreasonable to assume that 80 percent of the energy consumable by large industrial boilers can be displaced by gas and coal. The smaller boiler question is more difficult. However, using prior studies and taking advantage of the large proportion of natural gas used and recent developments in stoker-fired fluidized bed combustors and tri-fuel boilers, we show that it should be possible to displace most of the oil used by small industrial boilers.

Table 1.7 U.S. Energy Supply and Disposition

(Quadrillion Btu)

Activity and Fuel	1960	1965	1970	1975	1976	1977	1978	1979	1980[a]
Supply									
Production									
Crude Oil and Lease Condensate	14.93	16.52	20.40	17.73	17.26	17.45	18.43	18.10	18.25
Natural Gas Plant Liquids	1.46	1.88	2.51	2.37	2.33	2.33	2.25	2.29	2.27
Natural Gas[b]	12.66	15.78	21.67	19.64	19.48	19.57	19.49	20.08	19.70
Coal[c]	11.12	13.38	15.05	15.19	15.85	15.83	15.04	17.65	18.88
Nuclear Power	0.01	0.04	0.24	1.90	2.11	2.70	2.98	2.75	2.70
Hydropower	1.60	2.06	2.63	3.15	2.98	2.33	2.96	2.95	2.91
Other[d]	(l)	0.01	0.02	0.07	0.08	0.08	0.07	0.09	0.11
Total Production	41.78	49.67	62.51	60.06	60.09	60.29	61.20	63.91	64.82
Imports									
Crude Oil[e]	2.20	2.65	2.81	8.72	11.24	14.03	13.46	13.83	11.06
Refined Petroleum Products[f]	1.80	2.75	4.66	4.23	4.43	4.73	4.36	4.11	3.38
Natural Gas	0.16	0.47	0.85	0.98	0.99	1.04	0.99	1.30	1.03
Other[g]	0.07	0.04	0.07	0.19	0.18	0.30	0.44	0.39	0.28
Total Imports	4.23	5.92	8.39	14.11	16.84	20.09	19.26	19.62	15.75
Adjustments[h]	−0.44	−0.74	−1.41	−1.08	−0.21	−1.96	−0.36	−1.66	−0.54
Total Supply	45.57	54.85	69.49	73.09	76.72	78.43	80.10	81.87	80.03
Disposition									
Consumption									
Refined Petroleum Products[i]	19.92	23.25	29.52	32.73	35.17	37.12	37.97	37.12	34.25
Natural Gas[b]	12.39	15.77	21.79	19.95	20.35	19.93	20.00	20.67	20.44
Coal[e]	10.12	11.89	12.66	12.82	13.73	13.96	13.85	15.11	15.67
Nuclear Power	0.01	0.04	0.24	1.90	2.11	2.70	2.98	2.75	2.70
Hydropower[j]	1.65	2.06	2.65	3.22	3.07	2.51	3.16	3.17	3.13
Other[d]	(l)	0.01	0.02	0.07	0.08	0.08	0.07	0.09	0.11
Net Imports of Coal Coke	−0.01	−0.02	−0.06	0.01	(l)	0.02	0.13	0.07	−0.04
Total Consumption	44.08	52.99	66.83	70.71	74.51	76.33	78.15	78.97	76.27
Exports									
Coal[c]	1.02	1.38	1.94	1.79	1.62	1.47	1.10	1.78	2.47
Other[k]	0.46	0.48	0.73	0.60	0.59	0.63	0.85	1.12	1.29
Total Exports	1.48	1.86	2.66	2.39	2.21	2.10	1.95	2.90	3.76
Total Disposition	45.57	54.85	69.49	73.09	76.72	78.43	80.10	81.87	80.03

[a]Preliminary.
[b]Dry marketed gas.
[c]Includes bituminous coal, lignite, and anthracite.
[d]Geothermal, wood, refuse, and other vegetal fuels used for electricity generation at utilities.
[e]Includes imports of crude oil for the Strategic Petroleum Reserve.
[f]Also includes imports of unfinished oils and natural gas plant liquids.
[g]Includes bituminous coal, lignite, and anthracite, as well as coke made from coal, and hydropower.
[h]A balancing item. Includes stock changes, losses, gains, miscellaneous blending compounds, unaccounted for supply, and anthracite shipped overseas to U.S. Armed Forces.
[i]Refined petroleum products supplied includes natural gas plant liquids and crude oil burned as fuel.
[j]Includes industrial generation of hydropower and net electricity imports.
[k]Includes crude oil, refined petroleum products, natural gas, coke made from coal, and hydropower.
[l]Less than 0.005 quadrillion Btu.
Note: The sum of components may not equal the total due to independent rounding.

SOURCE: U.S. Department of Energy 1980 Annual Report to Congress DOE/EIA-0173 (80), March 1981.

The prospects for displacing oil in the process heat industrial sector are also quite good. In addition coal can be used directly as a chemical feedstock in place of oil. Furthermore biomass (Zaborsky 1981) and botanochemical crops can substitute for the use of many petroleum-derived products (Buchanan and Duke 1981).

Chapter 10 also contains a discussion that suggests that the U.S. might be able to make substantial inroads into the oil used in the commercial sector and the residential sector (at least for apartment houses and condominiums). Table 1.8 gives best, low and high estimates of oil displacement levels that might be achieved in the utility, industrial, commercial and residential sectors by vigorous programs directed toward gas-coal replacement of oil. Note that if the high estimates were realized we would practically have achieved National Energy Independence, the goal of our nation since 1974.

Finally, we should note that while concern for the United States' flagging economy has been the driving force behind this entire effort, the international aspects of oil-displacement by gas and coal should not go unnoticed. Most immediately Europe appears to be going ahead with a pipeline to draw upon natural gas from the U.S.S.R. Several countries in Europe also have large coal reserves. Many other countries of the world that are deficient in oil reserves have coal and natural gas. The economic and environmental advantages of gas-coal burning in place of oil would be accessible to such countries as well.

Table 1.8 National Potential for Oil Displacement in Million Barrels of Oil per Day

	Current Use	Best Estimate	Low	High
Utilities	1.4	1.0	0.8	1.2
Industry	3.4	2.0	1.6	2.6
Commercial	1.4	0.8	0.5	1.0
Residential	1.6	0.6	0.4	0.7
	7.8	4.4	3.3	5.5

1.9. Research and Development Needs

Chapter 11, Sections 1 and 2, briefly describes existing national research and development programs directed toward finding alternatives to oil and natural gas. More recently a general recognition is developing of a near term need to rely upon our abundant supply of coal (National Academy of Sciences, 1980; Landsberg, et al. 1980; ICAAS, CBI, 1980). A program to restore to coal use, boilers originally designed for coal but converted to oil, has for the most part been carried out. Conventional wisdom did not favor retrofitting to coal use boilers originally designed for oil or natural gas. However, with the retrofitting of two utility oil boilers at Kwinana in Australia to dual coal/oil capability, a more flexible view has developed and studies of various options for such oil to coal conversions are beginning. The favored order of research and development priorities for the displacement of oil in the United States appears to be coal-oil mixtures and after that coal-water mixtures. Gas-coal is not yet under serious consideration.

Since the analyses in this monograph suggest that the displacement of oil by burning coal with gas warrants serious consideration we devote the rest of Chapter 11 to a broad sketch of a Research and Development Program which would enable this country to implement this option quickly.

To provide a focus, Section 11.4 describes three conceptual approaches to oil boiler conversion to gas-coal burning. The first approach, direct gas-coal firing, retrofits the oil boiler to take the technical characteristics of coal boiler. However, unlike the Australian Kwinana solution, natural gas is used instead of the retention of oil boiler capability as a means of overcoming the derating to 60 percent of design capacity experienced with coal use alone.

The second approach introduces a first stage combustion chamber which has a multiple purpose: (1) to provide a combustible gas for afterburning in the boiler itself when augmented by natural gas; (2) to separate out most of the coal ash before the hot combustion products are injected into the boiler; (3) to provide for first stage SO_x and NO_x suppression and (4) to make the ash commercially useful.

The third configuration considers mixing gas with a coal-water slurry. Again a first stage combustion chamber is used to serve the functions defined in the second configuration and to facilitate the direct use of the larger particle sizes being considered for coal-water slurry pipelines.

These three configurations help define areas of uncertainty which should be resolved by a research program that is defined in Section 11.5. These include studies of: (1) the ash formation process in interacting gas-coal flames, (2) the characteristics of afterburning flames produced by various gas-coal primary flames, (3) the influence of additives in the primary stage upon the production of SO_x, NO_x and the possible commercial applications of the ash (4) the influence of the water in the gas-coal-water approach upon the ash formation process, the characteristics of the after burning flame, and the influence of additives. Also included in this section is a list of the principal research recommendations very recently formulated by an American Physical Society Study Group on Research and Planning for Coal Utilization and Synthetic Fuel Production (Cooper et al. 1981).

Using knowledge generated in the large coal-gasification program, by experience with cyclone burners and recent combustion studies on NO_x control, the immediate development of first stage gas-coal combustion chambers with ash separation and SO_x and NO_x suppression capability could be launched while basic and applied research are underway - a parallel rather than series approach. The broad outline of such a development program is defined in Section 11.6. Section 11.7 briefly describes interdisciplinary research needs.

Section 1.10   Final Summary

This effort began in January of 1981 shortly after the completion of two interdisciplinary assessments, Coal Burning Issues and Impact of Increased Coal Burning on Florida (ICAAS 1980a,b). It was motivated by a strong concern that we had overlooked the possibility that natural gas production rates could be sustained at a level of 20 TCF for the next two decades or so. As we looked into this question (Green and Jones, 1981), we became convinced that this possibility was appreciable. In February of 1981 we initiated a small development program and in March a scientific program. In June, we initiated an economic analyses of oil boiler conversion in Florida which was completed in August. The economic benefits indicated in the third endeavor (see Chapters 5-8 of this monograph) were far beyond our expectations and together with knowledge gained during the development and scientific programs these projected benefits convinced us that the gas-coal approach could and should openly stand as a contender with the coal-oil mixture, dual coal/oil, a new coal boiler and coal-water mixtures approaches to oil boiler retrofitting. The pay-off for the United States from oil boiler retrofitting is very large and we show in Chapter 10 that the displacement of 4 or so million barrels of oil per day seems quite feasible. This level of reduction would bring the United States within the range of a normal trade balance with the OPEC nations.

We must acknowledge that there are unusual obstacles to the implementation of the gas-coal approach. In the first place the Power Plant and Industrial Fuel Use Act of 1978, which assumed that gas production rates would decline precipitously by the year 2000, provides legal obstacles despite the fact that it is now being interpreted with very great flexibility. In the second place none of the agencies which would naturally support oil-backout research and development endeavors such as the Department of Energy, the Environmental Protection Agency, the Electric Power Research Institute, the Gas Research Institute or the National Coal Association has a preexisting program to pursue the gas-coal option. In days of very tight R and D budgets it is difficult to establish a new program. In the third place, to implement the gas-coal approach would require an unusual degree of cooperation by diverse groups including the electric utility industry, the boiler manufacturers, the gas and coal producing and distributing industries, and the federal and state regulatory agencies that are concerned with energy-environmental problems.

Despite these serious obstacles, we still believe, after a year of intensive involvement with scientific, technological, economic and environmental aspects of burning coal with gas, that the approach warrants serious consideration by the United States. Not only does it offer great promise as a way of backing out of using oil in existing boilers and furnaces but it also suggests a simple design path for improving the performance of new systems using a variety of solid fuels.

As 1981 draws to a close we feel that from a broad perspective we are basically close to where ICAAS was a year ago (see CBI and IICUF reports, ICAAS 1980a, b). To deal with our excessive oil imports problem we advocated strong conservation measures and increased reliance upon our nation's abundant supply of coal. However, among the ways of utilizing coal as an alternative to oil, we believe now that burning coal with gas is environmentally cleaner than coal alone. Furthermore it can permit coal use in boilers originally designed to burn oil without substantial derating. We hope that we have put together a reasoned presentation of various aspects of the gas-coal approach that will facilitate its early, or at least eventual, consideration.

In final conclusion we might note that during World War II this nation in a short span of time brought forth many new high technology developments - radar, pressurized high altitude aircraft, solid fuel rockets, and the release of nuclear energy, to name but a few examples. The technology to implement the displacement of oil by burning coal with gas does not compare in complexity with these wartime developments. However, since World War II we appear to have lost considerable economic strength and common purpose. Somehow, when addressing important national problems we must again set aside territorial imperatives, and political mistrust and work together towards the common good. Clearly we must strive towards a balance of trade which does not require that we sell our land and resources (i.e. our future) to make up for our immediate demands for foreign products, particularly oil.

CHAPTER 2

COAL AND NATURAL GAS SUPPLIES

2.1 Coal Supplies of the United States

Underlying the principal theses of this monograph is the assumption
that coal is an abundant resource for the United States and natural gas
is a viable additional energy source for at least the next two decades.
Extensive reviews have documented the availability of coal in the United
States. For convenience we present an overview of the coal supplies
using material borrowed from Coal Burning Issues (ICAAS, 1980) which
will be cited to as CBI.

In gross dimensions, the world's coal reserves are illustrated by
the data in Table 2.1 (CBI, p. 9) extracted from Peters and Schilling's
(1978) appraisal of world coal resources. The data are presented in
metric tons with the energy equivalence of hard coal (HC) rather than in
gross weight. Hard coal here encompasses anthracite and bituminous and
brown coal (BC) includes sub-bituminous and lignite. Gross supply data
also have been adjusted to reflect only supplies that are economically
and technically recoverable. Hence, other estimates of coal resources
could vary substantially from those shown. Be that as it may, this
table indicates that the United States has more recoverable coal resources than any other nation. The last two columns give the production and
export rates in 1975 in million tons.

Peters and Schilling also have predicted how rapidly coal will be
put into production in the coming years. As shown in Figure 2.1 (CBI,
p. 10), the United States is expected to lead the world by 1990. The
existing coal supplies of the USA are large enough, even if used up at a
voracious rate, to provide time for the development of even more
plenteous energy resources that may be developed, such as nuclear fusion
or breeder fission, or to learn how to live on renewable energy

23

Table 2.1 Economically and Technically Recoverable Coal Resources
of Leading Coal Countries in Gigatons (10^9 metric tons).
Production and Export are in 10^6 metric tons/yr (from CBI).

		HC	BC	Total	Prod	Exp
	World	492	144	636	2593	199
1	USA	133	64	177	581	60
2	USSR	83	27	110	614	26
3	China	99	---	---	349	3
4	Great Britain	45	---	45	129	2
5	West Germany	24	10	34	126	23
6	India	33	---	33	73	---
7	Australia	18	9	27	69	29
8	South Africa	27	0	27	69	3
9	Poland	20	1	21	181	39
10	Canada	9	1	10	23	12

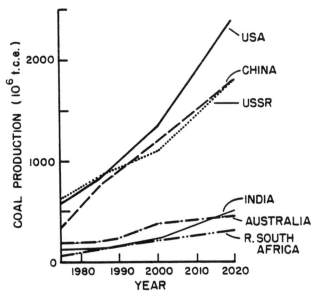

Fig. 2.1 Survey of the Future Trend in Production (CBI)

resources. At current production levels (~ 700 million tons per year) the estimated U.S. reserve of 177 gigatons of economically and technically recoverable coal would last 250 years. Even if the consumption rate were to triple, the supply would last nearly 100 years. This underscores an important purpose of this book; namely, to examine whether it is possible to make coal displace oil in existing oil boilers in an environmentally acceptable fashion by using natural gas as supplementary energy source and flame purifying agent.

In Table 2.2 (CBI, p36) the estimated identified resources by state to a depth of 3,000 feet are given for each of the three regions. These resources total 1,580 billion tons of which 685 billion tons are bituminous coal, 424 billion tons sub-bituminous coal, 450 billion tons lignite, and 21 billion tons anthracite. In addition to these known (but not necessarily all recoverable) resources, there are 1,643 billion tons of hypothetical resources (i.e., geologically predictable but undiscovered), for a total U.S. coal resource base of 3,223 billion tons, which is equivalent to the resource base of 71,000 quads, of which 4,933 quads or 225 billion tons of coal is presently considered recoverable. Of this amount, about 30 percent can be mined at the surface; the balance is underground.

Table 2.3 (CBI, p. 38) gives the distribution of U.S. coal by sulfur content, an important consideration for the environmental aspects of the gas-coal-oil displacement question.

2.2 Conventional Natural Gas Supplies

Because natural gas is an environmentally clean fuel, there is great interest on the part of many scientists and policy makers to assess its availability. Fortunately in the period 1980-1981, several estimates of reserves and probable production rates have been issued by major groups knowledgeable in the field. These estimates cover the period 1980-1981 to the year 2000. They range from very conservative to highly optimistic, if not enthusiastic. Almost all estimates take cognizance of the stimulating effects upon exploratory drilling of the partial deregulation of gas by the Natural Gas Policy Act (NGPA) of 1978. In the late seventies, very conservative estimates of resources

Table 2.2 Estimated Identified Coal Resources in the United States (Jan., 1972) (Million Short Tons) (Ohanian CBI)

	B[a]	SB[a]	L[a]	A[a]	Total
Eastern					
Alabama	13,342		2,000		15,342
Kentucky-East	32,421				32,421
North Carolina	110				110
Maryland	1,158				1,158
Ohio	41,358				41,358
Pennsylvania	56,759			20,510	77,269
Tennessee	2,572				2,572
Virginia	9,352			335	9,687
West Virginia	100,628				100,628
Total:	257,700		2,000	20,845	280,545
Central					
Arkansas	1,638		350	430	2,418
Illinois	139,124				139,124
Indiana	34,573				34,573
Iowa	6,509				6,509
Kansas	18,674				18,674
Kentucky-West	32,421				32,421
Michigan	205				205
Missouri	31,014				31,014
Oklahoma	3,281				3,281
Total:	267,439		350	430	268,219
Western					
Alaska	19,413	110,668			130,081
Arizona	21,246				21,246
Colorado	62,339	18,242		78	80,659
Montana	2,299	131,855	87,521		221,675
New Mexico	10,752	50,671		4	61,427
North Dakota			350,630		350,630
Oregon	50	284			334
South Dakota			2,031		2,031
Texas	6,048		6,824		12,872
Utah	23,541	180			23,721
Washington	1,867	4,190	117	5	6,179
Wyoming	12,705	107,951			120,656
Total:	160,260	424,041	447,123	87	1,031,511
Grand Total:	685,399	424,041	449,473	21,362	1,580,275

(a) B = Bituminous   (12,000 - 15,000 Btu/lb)
  SB = Sub-bituminous (8,000 - 11,000 Btu/lb)
  L = Lignite   (6,000 - 7,500 Btu/lb)
  A = Anthracite   (13,000 - 15,000 Btu/lb)

Table 2.3 Percentage Distribution of Coal by Sulfur Content (CBI)

	0 - 1.0%	1.0 - 2.0%	2.0 - 3.0%	3.0 - 4.0%	>4.0%
Eastern	5.0	4.5	5.3	1.5	-
Central	-	1.5	3.0	12.4	7.0
Western	57.0	2.8	-	-	-
Total	62.0	8.8	8.3	13.9	7.0

seemed to place natural gas as a less important source of energy for the United States. The current optimism of some experts arises from reassessment of resources, new technology for accessing methane below 15,000 feet, and large finds in the West, Oklahoma and Louisiana. There is also a strong opinion expressed by certain groups that the PIFUA legislation mandating coal use should be modified.

A review of several estimates of natural gas available between 1980 and 2000 reveals a considerable range of numbers and opinions. The Tennessee Gas Transmission Company (TGT 1980) in its previous publication Energy 1980-2000 (May, 1980) has taken a conservative approach. The view is presented that the natural gas production from the lower 48 states will decrease from a peak of 22.5 trillion cubic feet (TCF) available in 1972-1973 to approximately 10 TCF by the year 2000. In their publication (TGT 1981), the Tennessee Gas Transmission Company gives a lower 48 supply estimate of 12 TCF for the year 2000. However, the total United States gas supply may remain at the 1980 level of approximately 20 TCF through the year 2000. By 2000, approximately 40 percent of the total supply will come from supplemental sources such as Alaskan gas, liquefied natural gas (LNG), synthetic gas (SNG), etc. These estimates, however, require the "timely certification of supplemental projects . . ." (ibid., p. 16)

A very conservative view of natural gas availability for the future is presented in the Rand publication The Discovery of Significant Oil & Gas Fields in the United States (Nehring and Van Priest 1981). The report utilizes a data base consisting of all the significant oil and gas fields in the United States outside of the Appalachian region. The authors describe what has been discovered and within what time frame. They postulate why these discoveries were made and use their data base to assess the ultimate conventional oil and natural gas to be recovered in the United States. Using a historical analysis, Nehring and Van Priest provide a rather gloomy set of forecasts.

The authors estimate from their data base that the ultimate recovery of conventional natural gas in the United States will "most likely be between 920 and 1090 trillion cubic feet, as compared with a known recovery of 750 trillion cubic feet." (ibid., p. 5) They note that the number of important natural gas discoveries steadily declined from a peak in the 1950s with the exception of the activity in offshore Gulf of Mexico in the 1970s They suggest that the petroleum industry is gradually running out of ideas as to where oil and gas may be found (ibid., p. 135).

The data base developed by Nehring et al. uses December 1975 as its final date upon which forecasts are made. It estimates 751 TCF of known recovery, a range of 29-71-131 TCF for reserve growth; a range of 143-170-209 TCF for undiscovered reserves; and the range of 922-992-1091 TCF for ultimate recovery of natural gas in the United States (ibid., p. 176). The three values correspond to the 95 percent probability of more than that amount, and 50 percent and 10 percent probability estimates.

The report states that it is likely that more than half of all conventional gas reserves that will be ultimately produced in the United States have already been produced. In 1979 cumulative production equalled 578 TCF of natural gas (ibid., p. 170). Estimates are also given that only between 17 and 26 years of natural gas production remain at the 1979 levels (ibid., p. 171). While based upon an extensive data base and significant historical research, this Rand report has been criticized as being too pessimistic. Because of the 1975 cutoff dates, critics believe that economic changes and advances in exploration and drilling technology, particularly deep drilling technology, may not be adequately reflected in the estimates.

In October 1980, the American Gas Association (AGA) published a report on reserves and production of natural gas in the United States for the period 1980-2000 (AGA 1980). Its outlook is relatively optimistic. The report projects that sufficient quantities of methane will be available in the year 2000 to supply 25 percent of the total energy needs of the United States and possibly as much as 33 percent. It notes that while projections of supply can vary with hypotheses concerning future technical, social, and economic conditions, between 23 TCF to 33 TCF annual gas production may be available in the year 2000 (ibid., p. 2). The AGA Supply Committee emphasizes however, that an unprecedented capital investment will be required to make the supplies available. Moreover, sources of gas other than the conventional lower

48 production will be required. By the year 2000, supplemental sources may contribute from 40 percent to 60 percent of the total supply (ibid. p. 2). Although the AGA 1981 forecast for production in the year 2000 has not been published, their gas supply manager had indicated that it will be about the same as that given in the previous publication (Wingenroth 1981). An updated set of tables is given in Chapter 3.

In 1975 the U.S. Geological Survey published estimates of the undiscovered recoverable oil and gas resources of the United States. A new set of estimates was made in 1980 and published in 1981 (U.S.G.S. 1981). Reserves under the continental slopes are not included in these recent estimates. The report defines undiscovered recoverable resources as those which can be extracted economically under existing technology and price/cost relationships assuming normal short-term growth (ibid., p. 1). The offshore Alaskan assessment, however, includes some resources that are recoverable only if technology is developed to go beneath Arctic Ocean pack ice.

The assessments of undiscovered recoverable oil by the U.S.G.S. are presented in ranges of a low, 95 percent probability of more than that amount, a modal ("most likely") estimate, and a high, 5 percent probability of more than that amount. The results of this study indicate that the total conventionally producible undiscovered recoverable gas is estimated to be between 474.6 TCF (95%) to 607.0 (50%) to 739.3 TCF (5%) of gas (ibid., p. 5). The authors of the report point out that while exploratory drilling the Gulf of Alaska, southern California borderland, south Atlantic shelf, and eastern Gulf of Mexico has been disappointing, the drilling in the Cordilleran overthrust belt in the West has uncovered a large potential for gas. These data are reflected in the estimates (ibid. p. 7). These facts account for a more optimistic outlook of potential supply given in the 1981 report. No production estimates, however, are forecast.

The Potential Gas Agency's Potential Gas Committee estimated as of December 31, 1980, are reasonably optimistic (p. 6, 1981). The committee presents its potential conventional supply of natural gas in three types of estimates: probable, possible, and speculative. For total United States, there are 193-358-362 TCF respectively (ibid., p.

5). They also assume adequate economic incentives and that current or foreseeable technology will be utilized They do not, however, assume any fixed time schedule for discovery and production of natural gas. In the probable category of the lower 48 states they estimate 77 percent lies at drilling depths of less than 15,000 feet and 23 percent lies at drilling depth between 15,000 to 30,000 feet. Eighteen percent of the probable resource is offshore (ibid., p. 22). Finally, the estimate of ultimately recoverable conventional gas for the United States as of December 31, 1980, is lower than that made by the Potential Gas Committee in 1978. This is caused by transfers of gas from potential supply to produced reserves and production and increases and decreases in evaluation due to new data acquired. The area of major increases in estimates include the eastern Overthrust Belt, the deep Tuscaloosa trend, and other areas.

This brief review of estimates of conventional reserves and production of natural gas for the year 2000 has included only a few of those available. There are many others such as those produced by the Exxon Company, Mellon Institute, Mitre Corporation, U.S. Department of Energy, the U.S. Energy Information Administration and Southern State Energy Board (1981). The American Gas Association has also released a document comparing lower 48 production estimates for the year 2000 for a number of major oil producing companies, gas transmission companies, government agencies, and others. The figures range from 8.3 TCF to 30 TCF. For lower 48 production in 2000, Exxon forecasts 13.4 TCF; Standard Oil of Indiana, 15.3 TCF; the Energy Information Agency, 10.5 TCF; Mitre, 30 TCF; and the Department of Energy, 16.3 TCF (for 1990).

Most groups are more optimistic in the eighties in their estimates of gas reserves and production for the year 2000 than they were in the seventies. To some extent unconventional sources of natural gas, new technology, and new economic incentives, underlie this greater optimism. Nonconventional sources will be described in the next chapter.

CHAPTER 3

NONCONVENTIONAL SOURCES OF NATURAL GAS

3.1 Overview

Table 3.1 presents the December 1981 AGA projections of natural gas
supplies for various policy scenarios. Concentrating on the North
American focus we see that the 20 TCF production level should easily be
maintained. With a vigorous R&D effort the 30 TCF level should be
achievable in which case natural gas should be able to carry more that
the 25% share of the natural energy load which for the year 2000 is
estimated at 100 quad (~ 100 TCF). For these higher estimates
nonconventional sources play an important role.

Reviews of the status of nonconventional sources of natural gas
have been carried out by the Potential Gas Agency (PGA) (1981), the
American Gas Association (1981) and the Department of Energy (1981).
The sections of this chapter discuss nonconventional gas supplies as
follows: 3.2, geopressured-geothermal zones; 3.3, gaseous coal seams;
3.4, western tight sands; 3.5, (Devonian) shales; 3.6, gas hydrates;
3.7, abionic deep gas; 3.8, synthetic gas from biomass and wastes; 3.9,
synthetic natural gas (SNG) from coal gasification, and 3.10, liquefied
natural gas (LNG). The projected totality of all conventional and nonconventional natural gas sources depends significantly on the assumptions made. Figure 3.1 and Table 3.2 gives overviews of nonconventional
resource estimates. For perspective, conventional U.S. resource estimates lie in the 1000 - 2000 TCF range.

3.2 Geopressured Aquifers

The term geopressure is used to describe subsurface intervals where
the pressures encountered exceed those anticipated for a particular
depth. Under normal hydrostatic conditions, the fluid pressure for any

Table 3.1 American Gas Association Estimates (in Trillion Cubic Feet) from The Gas Energy Supply Outlook: 1980-2000 as Revised Dec. 1981

	1979	National Policy of Self Sufficiency		North American Focus		National Policy Accepts Moderate World Imports		World Conventional Gas Emphasis	
		1990	2000	1990	2000	1990	2000	1990	2000
Lower 48	19.9	16-17	13-14	15-17	12-14	15-17	12-14	15-17	12-14
SNG	0.2	0.3	0.3	0.3	0.3	0.3	0.3	0.1	0.1
Alaskan	---	1.3	2.8	1.3	2.8	0.8	1.4	1.3	2.8
Canadian	1.0	1.0	1.0	1.7	2.0	1.7	2.0	1.7	2.0
Mexican	---	0.1	0.1	1.0	2.0	1.0	1.5	1.0	2.0
LNG	0.2	0.7	0.7	0.7	0.7	0.8	1.3-2.5	1.0	2.5
Coal Gasification	---	0.7	2.7	0.7	2.7	0.2-0.7	1.3-2.5	0.2-0.7	1.3-2.5
Tight Formations	---	0.4-1.0	1.2-3.0	0.4-1.0	1.2-3.0	0.4-1.0	1.2-3.0	0.4-1.0	1.2-3.0
Nonconventional	---	0.2-0.6	1.0-2.5	0.2-0.6	1.0-2.5	0.2-0.8	1.0-2.5	0.2-0.8	1.0-2.5
Total Supply	21.3	20.7-22.7	22.8-27.1	21.3-24.3	24.7-30.0	20.4-24.1	22.7-29.2	20.4-24.6	24.9-31.4

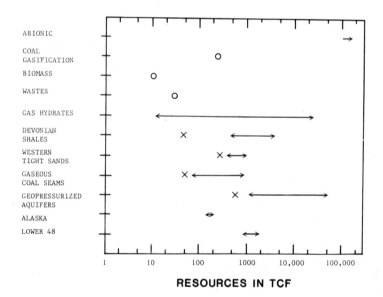

RESOURCES IN TCF

Fig. 3.1 Natural Gas Reserve Estimates Associated with Conventional and Nonconventional Sources. The x's Denote Recoverable Estimates. The Circles Denote 100 Times the Predicted Annual Rate for the Year 2000 A.D.

Table 3.2 Nonconventional Natural Gas Supply Totals-In-Place According to Various Sources Given in the Potential Gas Committee Report 1981

Geopressured Aquifers		Gaseous Coal Seams		Western Tight Sands		Devonian Shales		Gas Hydrates	
Ref	Total in Place	Ref	Total in Place	Ref	Total in Place	Ref	Total in Place	Ref	Total in Place
W-79	5,700	K-80	550	NPC-78	924	NKM-80	2,995	T-77	9,392
L-78	1,100	NPC-80	398	L-78	709	FER-79	285	MC-79	27.000
D-77	5,735	S-80	500	NGS-78	793	S-78	431	D-79	6.7×10^6
J-77	49,000	R-79	249	NGS-73	600	UGS-76	548		
H-76	3,000	NGS-78	505	R-78	100				
P-75	23,618	O-78	494						

Table 3.3 Sources of Biomass

Source	Commonly Mentioned Varieties
Aquatic Biomass	
● Marine Farming	Giant kelp, macroalgae
● Onshore Aquaculture	Water hyacinths, algae, duckweed, water pennewort
Land Biomass	
● Silviculture	Pine, hardwoods, eucalyptus, cull trees, scrub trees
● Grasses and Crops	Kudzu, alfalfa, corn, sorghum, sugarcane, sugarbeets, potatoes, kenof, cassava, rice
● Biomass Residues	Wastes from forestry operations, wood processing, and crop harvesting and processing

subsurface point may be approximated by applying a gradient of 0.46 psi per foot of depth, which is roughly equivalent to the weight of a column of normal marine water of the same height. With geopressured zones the pressure gradient may be as great as 0.98 psi/foot, largely reflecting the weight of the overlying sedimentary rocks. Geopressured zones have been found in various regions of the United States. The largest and most likely commercial targets are located in the Gulf Coast region. In general, geopressured zones arise from the compaction of rapidly deposited clays and sands of low overall permeability. Such conditions are commonly associated with deltaic sedimentation which along the Gulf Coast attain thicknesses of 50,000 feet. The quick deposition of these sediments resulted in clay compaction and the isolation of porous and permeable sand with impermeable shales. In addition, large growth faults developed which compartmentalized these sands.

Geopressured aquifers in the Gulf Coast constitute a large natural gas reserve estimated to be from 1100 to 49,000 TCF. The first column of Table 3.2 gives geometric mean values from various sources given in PGA (1981). The estimated recoverable resources fall between 50 and 1200 TCF. The time for tapping this resource depends less upon new technology than upon sociolegal and environmental uncertainties.

3.3 Gaseous Coal Seams

Methane is a by-product formed during the conversion of organic material to coal, a process referred to as coalification. Though much of the gas is lost to the atmosphere, a substantial portion is retained as (a) gas absorbed on the surface of the microporous structure of the coal, (b) free gas in vertical joints and fractures (cleats) in the coalbed, and as (c) desorbed gas trapped within adjacent sediments. Classes (a) and (b) represent a nonconventional gas resource that is directly determined by the nature of the producing interval, namely the coal seam. Class (c) represents a resource that is difficult to classify as a conventional or nonconventional resource. In situations where adjacent sediments are porous and permeable, trapped gas derived from coal seams can be produced by conventional drilling and completion procedures.

Because of the natural association of methane with coalbeds, the geological dimensions of the gas resource are defined by the extent of the subsurface coalbeds. The amount of gas to be found within a given coal unit is a function of the rank of coal and the depth of burial. The rank of coal is important since methane content is inversely proportional to the moisture of the coal which decreases with an increase in coal rank. Depth of burial is important because it determines the pressure of the coal seam, and gas content increases with an increase in pressure. The range of estimated total U.S in place gas is 72-86 TCF. Table 3.2 provides individual estimate for gaseous coal seams with the geometric mean used wherever a wide spread is given.

Recovery of 40-60 TCF appears very feasible. The ultimate recovery will depend upon the solution of a variety of sociolegal and technological problems.

3.4 Western Tight Sands

The "tight" formations in the West occur in two types of formations, blanket and lenticular sands (American Gas Association 1980). Blanket sands, at one extreme, are single, relatively thin (10 to 100-foot) gas-bearing zones of generally uniform thickness that extend over a large area. At the other extreme are the lenticular formations, composed of relatively thick sections (possibly 1000 or more feet) scattered throughout the section, as in the nonmarine formations of the Rocky Mountain basins. There are 20 known geologic basins that have reservoir rocks with properties that are characteristics of unconventional sources: with low permability and less than 10 percent porosity. The comparison of estimated total U.S. in place tight-gas resources ranges from 409-924 TCF as shown in Table 3.2, third column.

Between 190 and 570 TCF may ultimately be recovered given reasonable prices and normal improvements in technology.

3.5 Devonian Shale

Devonian natural gas pockets are produced from the fractures, micropores, and bedding planes of shales deposited during the Paleozoic Devonian Period. The brown shales of the Appalachian Basin lie in a thick wedge of marine and terrestrial rocks that make up the large Catskill Delta complex. These shales of the Appalachian gas-well drillers' logs are black and brown, containing as much as five to 25 percent kerogen.

The Devonian brown shales are the reservoir rocks and presumably the source rocks in the Big Sandy Gas Field of southeastern Kentucky and adjacent West Virginia. These brown shales underlie about 160,000 square miles of the western and larger segment of the Appalachian Basin. The total volume of the brown shales is 12,600 cubic miles as determined from a study of well cuttings.

Various estimates of Devonian shale resource range from 285-3900 TCF with geometric means shown in Table 3.2. These are based upon the assumption of 0.23 to 0.41 cubic feet of gas per cubic foot of known shale. The greatest hindrance to recovery of this resource is the marginally favorable economic environment. In part this is a reflection of the low permeabilities and modest production rates associated with shale wells.

3.6 Gas Hydrates

Gas hydrates are an ice-like mixture of gas and water, a type of clathrate, in which gas molecules are trapped by a framework or "cage" of water molecules. Gas hydrates form in a gas-fresh water system under univariant pressure/temperature conditions such that, depending on the pressure, hydrates can be stable from subzero temperatures to more than 30^0C. The presence of salt (NaCl) decreases the hydrate stability field.

Potentially, large quantities of natural gas (either biogenic or thermal methane plus ethane, propane and other light hydrocarbons) can be trapped in gas hydrates. Also, large quantities of free gas or other hydrocarbons can be trapped beneath impermeable gas hydrate layers. An ideal methane-saturated fresh-water system at standard temperature and pressure can yield free gas at approximately 132 ft^3/ft^3.

Theoretically, the proper conditions (high pressure, low temperature) for gas hydrate formation exist in permafrost areas and within marine seafloor sediments. Estimates of the gas hydrate resource base are highly speculative and range from 500 to 1,200,000 TCF for permafrost areas and from 110,000 to 270,000,000 TCF for ocean sediments. From estimates of

the United States permafrost areas the gas resource estimates range from
11 to 25,000 TCF. For submarine sediments   the gas resource
estimates range from 2,700 to 6,700,000 TCF as in last column of Table 3.2.

3.7 Deep Abionic Gas

An enthusiastically optimistic view of the potential supply and
production of natural gas is projected by Russian (Kropotkim and Valiaev,
1976) and American (Gold and Soter, 1979, 1980) advocates of the deep-
earth gas hypothesis. The scientific (Gold and Soter 1979) and popular
(Gold and Soter 1980) articles describe the deep-earth origins of methane,
its source, and its path from the depths of the earth up into the earth's
atmosphere. The basic sources, they postulate, are not bionic but rather
the primoridal hydrocarbons present in the formation of the solar
system. Using this theory, discoveries of methane would require a
different strategy for locating reserves. The deep-earth gas hypothesis
has also been considered by the Russian geophysicists who began to evolve
their theory in the 1890s. Of course, a basic corollary of the theory is
that deep drilling should uncover a portion of these massive methane
resources. However, deep gas (i.e. gas below 15,000 feet) also has a more
conventional explanation (Oppenheimer 1980). If the abionic deep gas gas
hypothesis were correct, we would have an unlimited supply of natural gas
available if the technology of tapping it could be developed

3.8 Gas from Biomass and Waste

Biomass, in a universal sense, includes all growing plant life. In
the context of a supplement fossil fuel source, biomass includes aquatic
and terrestrial plants such as those listed in Table 3.3 which indicates
some sources and commonly mentioned varieties of biomass.

As yet, there is no commercial gas from biomass. Current research
efforts relating to biomass are extensive and broad based. The

University of Florida and the Gas Research Institute have recently launched a major project directed towards the production of methane by biological processes (Smith et al. 1979). These research activities have led to the initial isolation of many of the now known species of methane producing bacteria. Fermentation studies are underway which should lead to better control and regulation of the process and it is believed that a significant production level eventually can be achieved.

Pyrolysis of biomass can directly produce fuel gas on a fast time scale (Eoff 1979). A considerable background literature exists on this topic but only recently has economic urgency again directed attention toward this method of conversion.

Methane generation from human, animal, and agricultural wastes was a favorite topic among "back to nature persons" a dozen years ago. With the recent energy crisis this area is now receiving official attention (National Academy of Sciences 1977).

Urban refuse is the accumulation of garbage that is generated each day by society. Also known as municipal solid waste (MSW), it includes household and commerical waste, but not industrial, agricultural, construction, and sewage waste. Urban refuse is 75 percent organic matter. Much of the remaining 25 percent, consisting mainly of glass, metals, and ceramics, can be recovered and recycled.

Using EPA estimates, approximately 200 million tons of organic municipal solid waste will be generated in the year 2000. About 60 percent of generated MSW is collectible. An estimate of collected organic matter in the year 2000 would therefore be 120 million tons. The techniques of converting these wastes to methane are similar to those currently under study in the pyrolysis of biomass. The potential is quite significant with 0.11 TCF estimated for biomass and 0.3 TCF for solid waste by AD 2000 (AGA 1980, Tables X1 and X2).

3.9 Gas from Coal Gasification

Coal gasification constitute an important fall-back technology for the continued use of gas-coal burning in oil boilers in the event conventional sources are depleted faster than expected and other nonconventional sources do not fulfill expectations. An extensive technology background exists on this topic (Miller and Lee, Chapter 6, CBI). Table 3.4 summarizes gasification processes which have reached the pilot plant stage of development. How far we get with coal gasification is more a matter of economics and public policy than technology. Among the advantages of coal gasification is that on an end-use energy equivalency basis, high-BTU coal gasification (including gas combustion) would produce somewhat less CO_2 emissions to the atmosphere than equivalent coal fired electric power generation. Further, much of the CO_2 produced by coal gas plants would be highly pure and suitable for injection for enhanced oil recovery (Krickenberger and Lubore 1981).

3.10 Liquefied Natural Gas

Liquefied natural gas (LNG) in reality is conventional natural gas liquefied to -260^0F at atmospheric pressure. At this temperature its density is 600 times that of the vapor which permits it to be stored and tranported in large quantities. The advantages of LNG are that it would permit this country to import supplies of natural gas from countries too far away to make deliveries by pipeline. The extent of use of LNG is also more a matter of policy than technology. Since the gas-coal approach derives most of its energy from coal it would not be unreasonable to use imported liquid natural gas, to facilitate the clean use of domestic coal.

Table 3.4 Gasification Processes

Process Name	Developer (Reference)*	State of Development	Reactor Type	Temp (°F)	Pressures	Feed Gas	Products %b	HHV, Btu/ft^3
Bigas	Bituminous Coal Research (Grace, 1975)	120t/day pilot plant	Ent.	2400-2700	100-1500 psig	St^a, O$_2$	8	380
C-E	Combustion Engineering (Winterson, 1975)	5t/day pilot plant operating	Ent.	3200	1 atm	St, Air	<5	120
CO2-Acceptor	Consolidation Coal Co. (Fink et al., 1975)	40t/day pilot plant operating	Flu.	1500-1900	150-300 psig	St	17	440
GE Gas	General Electric (Kydd, 1975)	3/4t/day pilot plant operating	Stir.	N.A.	0-300 psig	St, Air	<5	160
Hybrid	Hitachi Research, Japan (Miyadera et al., 1978)	Laboratory scale unit	Flu.	1700	70-300 psig	St, O$_2$	18 T	460
Hydrane	U.S. Bureau of Mines (Gray and Yovorsky, 1975)	24t/day pilot plant planned	Flu.	1500	1000-1500 psig	St	73	820
HyGas	Institute of Gas Technology (Anastasia and Bair, 1975)	75t/day pilot plant	Flu.	1750	1000 psig	St, H$_2$	20 C	370-550
Kiln Gas	Allis-Chalmers (Chem. and Eng. News, 1978)	10t/day pilot plant in construction	Tum.	800-1000	200-500 psig	St, Air	5	100-200
Koppers-Totzek	Heinrich Koppers (Whiteacre et al., 1975)	15 plants in commercial operation	Ent.	2750-3300	1 atm	St, O$_2$	0	300
Lurgi	Lurgi Mineralotecknik (Bodle and Vyas, 1975)	14 plants in commercial operation	Mov.	1150-1400	300-500 psig	St, O$_2$	5 TH	300
Molten Iron	Applied Technology, Inc. (LaRosa and McCarvey, 1975)	Laboratory scale unit operating	Molt.	2500	1 atm	St, Air	0	160
Molten Salt	Atomics International (Kohl et al., 1978)	1t/day pilot plant in construction	Molt.	1700	1200 psig	St, Air	0	150
Synthane	U.S. Bureau of Mines (Haynes and Forney, 1975)	70t/day pilot plant	Flu.	1800	500-1000 psig	St, O$_2$	15 TC	400
Texaco	Texaco (Hottell and Howard, 1971)	Pilot plant operating	Ent.	2000	1 atm	St, Air	<5	170
Tri-Gas	Bituminous Coal Research (Colaluca et al., 1979)	Laboratory unit in operation	Flu.	1000	250 psig	St, Air	<5	150
U Gas	Institute of Gas Technology (Patel and Loeding, 1975)	Demonstration unit in design	Flu.	1900	350 psig	St, Air	4	150
Westinghouse	Westinghouse (Andermann, 1978)	16t/day pilot plant operating	Flu.	1800-2000	250 psig	St, Air	0	140
Winkler	Davy Powergas, Inc. (Banchik, 1975)	16 plants in commercial operation	Flu.	1500-1850	15 psig	St, O$_2$	2	270

Ent. - entrained bed; Flu. - fluidized bed; Stir. - stirred bed; Tum. - tumbling bed; Mov. - moving bed; Molt. - molten bath

a St - steam   b % methane   T - tars   C - char   TH - Tar, heavy oil   TC - tar, char;   N.A. - information not available

* See Coal Burning Issues, University Presses of Florida, 1980, for complete references.

CHAPTER 4

THE PHYSICAL BASIS FOR BURNING GAS-COAL IN PLACE OF OIL

The science of combustion is a highly complex combination of
physics, chemistry, fluid mechanics and other science and engineering
disciplines. Because of this complexity our understanding of the
combustion of the simplest fuels with air is far from complete. Our
understanding of the combustion of mixed fuels is at a more primitive
stage although there is considerable experience with gas-oil mixing and
more recently with coal-oil mixing. This work is concerned with the
potential application of burning of gas-coal mixtures to simulate oil in
boilers originally designed for oil for which there is very little
experience and practically no scientific analyses. In approaching this
topic it would therefore be best to start with comparative descriptions
of burning of the three major fuels used in utility boilers: natural
gas, residual oil, and bituminous coal.

4.1 Gross Comparative Features of Gas, Oil and Coal

Natural gas, a gaseous hydrocarbon fuel consisting primarily of
methane, is a highly attractive fuel. Transportation by pipeline and
handling are simple, and local storage is not required. Burners are
very simple, almost like a garden spray nozzle, and there are no ash
and SO_x problems. Unfortunately, there are regulatory problems
associated with natural gas. These stem from pre-1978 estimates of
natural gas reserves, which suggested that annual U.S. production would
decline from about 20 trillion cubic feet (TCF) to about 10 TCF by the
year 2000. Based on this assumption, the Powerplant and Industrial Fuel
Use Act (PIFUA) of 1978 was designed to phase out the power plant and
industrial use of oil and natural gas. However, as shown in Chapters 2
and 3, recent resources and production data do not support projections

41

of declining natural gas production. Indeed, most projections now suggest a continuous annual production capability to the year A.D. 2000 of about 20 TCF. Optimistic projections suggest that even 30 TCF might be achieved. In the present analysis we will accept the middle-of-the-road projection, that gas can provide approximately its present proportion, 25 percent of the energy needs of the country through the year A.D. 2000.

Oil, a liquid hydrocarbon, is a more neutral fuel. Handling and storing requires reasonable technologies. However, residual oil - the predominant grade of oil used in steam boilers - is a relatively dirty, asphalt-like substance, the burning of which entails problems of controlling SO_x, NO_x, heavy metals, and some ash. Until now the United States could not produce enough oil to satisfy its needs for transportation, for which the liquid form of fossil fuel is essential at this time. Recent developments have made it possible to convert residual oil into higher hydrocarbons such as gasoline, diesel fuel, and the like so that it may soon be possible for the U.S. to provide enough feedstock for all of its fuel needs for transportation. The continued use of the residual oil by utilities and industry, therefore, can only be viewed as fostering our excessive reliance upon OPEC oil and contributing to the severe economic problems of the United States that are associated with importing costly oil.

Coal, a solid hydrocarbon might figuratively be viewed as a repulsive fuel although it must be recognized at the outset that there is a broad range of coal qualities. The technology of coal burning is highly complex, requiring elaborate facilities for transportation and storage, for grinding and pulverizing, for delivery of coal to the burners, for environmental control of bottom ash. fly ash, and other suspended particulates, and for SO_x and NO_x suppression. Yet for the United States, coal has an overpowering redeeming feature--it is in goodly supply. Indeed, as seen in Chapter 2, the United States could be referred to as the Middle-East of coal.

Granting the abundant availability of coal, a fuel figuratively with repulsive characteristics (high pollution potential, lower power density) and the moderate availability of natural gas, a fuel with attractive characteristics (low pollution potential, high power density), we propose to burn coal simultaneously with natural gas in oil boilers. The intent is to use the attractive features of gas to cancel some of the repulsive features of coal to achieve the more neutral qualities of oil (accepted level of pollution, middle power density).

This figurative cancelation is analogous to the explanation of the basic nuclear force, one of the most difficult and time consuming problems of modern science. This is given in terms of the forces due to a mixture of meson fields. A vector meson field (like coal fuel) generates a strong repulsion whereas a scalar meson field (like gas fuel) contributes a strong attraction. When both fields are present the main terms cancel (Green 1970) and the small residual forces and minor ingredients do exceedingly well (like oil) in explaining experiments (Green, MacGregor and Wilson 1967, Measday et al. 1978).

To pursue details of the gas-coal-oil displacement approach it would be helpful to go further into the combustion of natural gas, residual oil and bituminous coal.

4.2 Natural Gas Burning

Natural gas, a gaseous hydrocarbon, as delivered by a pipeline consists primarily ($\sim$ 90 percent) of methane (CH_4), secondarily ($\sim$ 6 percent) of ethane (C_2H_6) and tertiarilly ($\sim$ 2 percent) of propane (C_3H_8). The residual gases including nitrogen, higher hydrocarbons, carbon dioxide, etc. constitute about 2 percent.

The combustion of methane, while by far the simplest of the three major boiler fuels, is still not well understood. Elementary chemistry would suggest that it proceeds via the reaction

$$CH_4 + 2O_2 \rightarrow CO_2 + 2H_2O$$

In point of fact many intermediate species play an important role in the overall combustion process and advanced works on combustion cite many simultaneous reaction mechanisms. For example, the chain of reactions given in Table 4.1 represents one of the simpler systems of reactions proposed to explain the burning of methane (Kanury, 1977).

Table 4.1 Methane Reactions

$$CH_4 + OH \rightarrow CH_3 + H_2O \qquad CO + OH \rightarrow CO_2 + H$$
$$CH_4 + O \rightarrow CH_3 + OH \qquad H + O_2 \rightarrow OH + O$$
$$CH_3 + O_2 \rightarrow H_2CO + OH \qquad H + H_2O \rightarrow OH + H_2$$
$$H_2CO + OH \rightarrow HCO + H_2O \qquad O + O + M \rightarrow O_2 + M^*$$
$$HCO + OH \rightarrow CO + H_2O \qquad 2OH + M \rightarrow H_2O + O + M^*$$

A more complex system of 41 reactions involving 10 intermediate species H, O, OH, CO, H_2, CHO, HO_2, CH_2, CH_3, CH_2O is given by Thurgood and Smoot (1979). Yet even this system neglects some possible intermediates. When consideration is given to the fact that air is usually the oxidizing medium then N_2 and various combinations of N with H, C and O also participate in the reaction system. At normal combustion temperatures these are at trace levels to extents which depends upon the temperature, pressure and fuel/air mixture ratio.

While nothing is really simple in the science of combustion, the burning of methane or other gaseous fuels does have the simplicity that the molecular form of the fuel permits it to directly undergo chemical reactions. This occurs if the fuel air mixture lies between the so-called flamability limits and if a sufficiently energetic local heat source such as a spark, pilot flame, hot electrical filament, or the combustion zone of a boiler ignites the mixture. Burners used to feed natural gas into a furnace take on a variety of forms (Babcock and Wilcox, 1978, p 1.5, to be referred to as BW). Individual spuds are usually arrayed in a circle at distances with respect to the burner throat which can be varied for optimum firing conditions. The gas spuds drillings are arranged to achieve flame stability. Natural gas fed through the spud drillings are rapidly and intimately mixed with combustion air flowing through the burner throat. The flame which is easily controllable is free of ash and SO_2.

4.3 Residual Oil Burning

The science and engineering of the combustion of residual oil, the main form of oil used in utility boilers, is more complex. Residual oil, a by-product of oil refining, has a heating value of about 18,300 BTU/lb (135,000 BTU/gal). It has a very high viscosity which requires that it be raised to about 250°F to pump and spray into the furnace. The burner produces a fine mist, somewhat like a heavy fog (BW, ibid. p. 7.4). High pressure steam or air "atomizers" are used with oil pressures up to 300 psi and steam pressures up to 150 psi. These burners are now available in sizes up to 300 million BTU/hr input, which corresponds to about 16,500 lbs of oil per hour (BW, ibid. p. 7.4). Typically oil droplets are about 50 microns (50×10^{-6} meters) in diameter which is large compared to atomic or molecular dimensions. Accordingly, the actual burning process on a microscopic scale must proceed through several intermediate steps.

Each liquid droplet when injected into the combustion zone of the boiler evaporates at the droplet surface. The fuel vapor diffuses outward mixing with oxygen diffusing inward. Thus chemical reactions occur in the vapor phase. The complexity of these reactions is undoubtedly greater than the methane reactions but few attempts have been made to model the oil reaction processes in such detail. What happens empirically is that a shell of flame surrounds the oil droplet which gradually shrinks in size as the droplet evaporates. The burnout time to complete evaporation is proportional to the square of the original droplet diameter (Kanury 1980 p. 167, Glassman p. 259). Thus to speed up the fundamental combustion process requires that the fuel droplets be made as small as possible. However, it takes energy to produce a fine mist so some engineering tradeoff is used in designing the boiler-burner fuel delivery system.

4.4 Coal Burning

The word "coal" encompasses a great variety of solid fuels which are primarily classified by ranks (Babcock and Wilcox (BW) 1978, Combustion Engineering (CE) 1981, Smoot and Pratt 1979, CBI 1980). Here we will mainly be concerned with bituminous coal which by far is the

largest available group and the most extensively used in utility
boilers. This coal derives its name from the fact that upon heating it
often reduces to a cohesive binding sticky mass. Almost all large
modern coal fired boilers utilize pulverized coal because pulverized
coal burns approximately like oil or gas. The coal is first crushed to
pass through a sieve with 1 1/4" to 1 1/2" openings. Then a variety of
pulverizers: ball-tube mills, role and race, ball-race, roller, or
impact, or hammer mills (CE Chap. 12, BW Chapter 9) are used to reduce
the coal to fine particle sizes. Typically pulverized coal is brought
to a fineness such that 70 to 90 percent passes through a 200 mesh
screen. This corresponds to the fineness of face powder or an average
particle size of about 50 microns. From the combustion viewpoint it
would be better if the particles were finer yet; however, the energy and
equipment costs of pulverizing coal go up rapidly as the particle sizes
go down.

At a microscopic level the burning of a coal particle is much more
complicated than the burning of an oil droplet which as we have seen is
more complicated that the burning of methane molecules. As we have
already discussed, methane combustion itself is complex by virtue of the
chain of reactions involved. When injected into a combustion zone the
coal particle first undergoes thermal decomposition usually referred to
as pyrolysis: "death by heat". Entrained volatiles and water are first
driven out of the interior. The combustible volatiles which include CH_4
and H_2 burn in air in a similar way to the liquid oil droplet producing
a flame shell surrounding the particle. Once the volatiles are
exhausted, the surface of the carbonaceous solid residue is attacked by
inwardly diffusing oxygen producing an incandescent surface. This
surface reaction of oxygen and carbon releases carbon monoxide which
migrates outward and upon encountering oxygen burns to carbon dioxide.
A weakly visible blueish envelope of flame thus surrounds the solid
particle as it burns out. Theories of the time and radial dependence of
the temperature and concentrations of CO, CO_2, and O have been developed
(See Kanury 1977, Glassman 1977, Wolfhard et al.. 1964) that are roughly
in accord with observations by Coffin and Browkaw (1957) and Hottel and
Stewart (1940). Burnout time again is proportional to the square of the

initial particle diameter (Kanury 1977, Spaulding 1951) but the constant is usually about 10 ± 5 times larger than for a liquid droplet depending upon the amount of volatiles originally contained.

As the coal particle is pyrolyzed and burned, it releases an ash residue which has many characteristic forms depending on the mineral substances originally entrained in the coal. The amount of ash usually varies between 5 to 20 percent of the original weight of the coal. This ash presents complex problems to the design and operation of coal boilers (and more complex problems yet to the retrofitting of an oil boiler for coal burning). Thus both the technology of coal burning and the microscopic scientific description of coal particle combustion are highly complex. Let us next examine morphologically the possibility that burning a mixture of coal and gas might be used to displace oil in an oil boiler.

4.5 The Combustion of Gas and Coal Mixtures

Successful exploratory tests of burning various mixtures of gas with coal (Green and Green 1981, Green et al. 1981), several literature citations on gas-coal firing arrangements used in the cement, lime, and brick industries (Zinn 1965; Brust 1978, Vunkov 1961), and experience with several utility boilers originally designed for coal with provisions for gas firing (Plank, 1981) indicate that gas and coal can be burned together in a variety of arrangements. In addition several quantitative considerations favor the idea that a combination of gas and coal could replace oil in oil boilers.

The first consideration favoring gas-coal burning is the power density argument. This is illustrated by Figure 4.1 which shows the relative volumes of typical boilers with the same power ratings (see also Fig 1.4). The first three figures from the left indicate that the relative power densities achieved in gas:oil:good coal boilers vary as about 1.60:1:0.73. This suggests that gas and coal burning in about 1:2 proportions by energy content could be carried out at the normal power density of oil burning.

A second consideration is the fuel calorific value per unit weight. Residual fuel oil typically yield about 18,000 BTU/lb whereas a

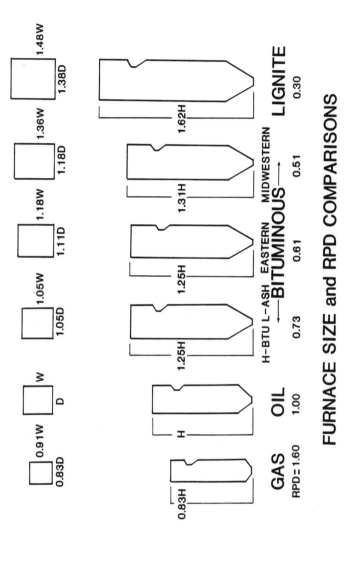

Fig. 4.1 Furnace Size and Relative Power Density Comparisons Relative to an Oil Boiler.
(adapted from Bogot and Sherrill 1976, Breen and Sotter 1978, Combustion Engineering 1981)

high grade coal yields about 14,000 BTU/lb and methane about 24,000 BTU/lb. Thus burning 60 percent by weight of coal together with 40 percent by weight of natural gas should lead to 18,000 BTU/lb, the same calorific release per pound as residual oil.

The hydrogen to carbon weight ratio provides another simple quantitative argument favoring the use of natural gas and coal to replace oil. The hydrogen to carbon weight ratio of residual oil is approximately 1 to 9. Methane (CH_4) is 4/16 = 1/4 hydrogen and 12/16 = 3/4 carbon by weight. If coal were predominantly carbon a burning proportion of 2 carbon atoms to 1 methane molecule, about 60 percent by weight of coal, and 40 percent by weight of gas will duplicate the hydrogen:carbon weight ratio of residual fuel oil. In actuality the greater the proportion of coal volatiles the less is the proportion of natural gas needed.

The hydrogen-to-carbon atom ratio is perhaps a better index. This is illustrated in Figure 4.2 which shows the hydrogen-to-carbon atom ratio of various hydrocarbons. Note that residual oil is approximately 1.6 whereas coals vary around 0.8. On the other hand pipeline natural gas consists primarily of methane (CH_4), secondarily of ethane (C_2H_6), thirdly of propane (C_3H_8) with residual gases such as nitrogen, oxygen and higher hydrocarbons totaling but a few percent. It follows by a simple calculation that about 60 percent by weight of coal, and 40 percent by weight of natural gas will duplicate the hydrogen:carbon atom ratio of residual fuel oil.

The air/fuel ratio on a weight basis also suggests the intermediate character of oil between natural gas and coal. This ratio typically takes on the values 20, 17, and 11 when burning gas, oil and coal respectively.

One of the most important aspects favoring the gas-coal displacement of oil is the fact that oil as a liquid is intermediate in its physical state between methane - a gas and coal - a solid. To burn oil at the high rates required by utility boilers it is necessary to first "atomize" the oil and disperse it into the furnace as a fine mist with droplets of about 50 microns size. To burn coal at high power rates needed in modern utility boilers it is first necessary to pulverize the

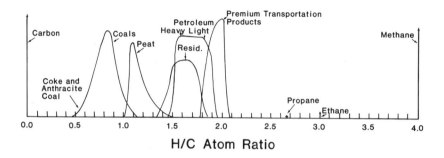

Fig. 4.2 Hydrogen/Carbon Ratio for Various Hydrocarbons (Adapted and
Extended Whitehurst 1978, see also Cooper et al. 1981).

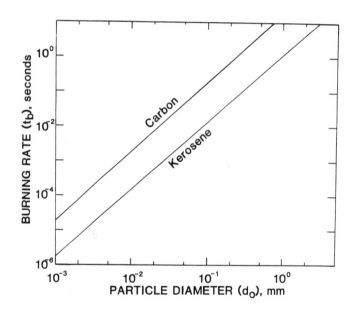

Fig. 4.3 Burning Rate vs. Particle Diameter for Carbon and Kerosene
Particles (Adapted from Kanury 1975).

coal into a fine powder with particle sizes also, distributed about the
50 micron size range. Figure 4.3 shows the burnout time for carbon and
kerosene particles of various diameters which illustrates the liquid-
solid burnout time difference. It has been noted that if the volatile
content of coal is low a coal line on Figure 4.3 would approaches the
carbon line. If the volatile content is high the coal line would
approach the liquid line.

The burning of a natural gas does not involve the particle burnout
step at all since the gas is already atomized (really molecularized)
upon injection into the furnace and is already in the gaseous state.
The burning rate of a mixture of gas and coal might be expected to
approach that of oil just as the burning rate of high volatile coal
approaches the liquid fuel line. In addition the local heating of the
coal particles by the burning natural gas might be expected to speed up
the pyrolysis of the coal particle and to accelerate all reaction
rates. It is well known in kinetic theory that reaction rates increase
rapidly with increases in temperature. All in all from a combustion
standpoint a mixture of gas and coal should burn somewhat like oil.

4.6 Environmental Features and Ash

Sulfur dioxide pollution considerations also indicate that a mixture of gas and coal can simulate oil. Let us take as a representative
example, a residual oil which yields 1.5σ where σ here denotes lbs
SO_2/million BTU. Let us choose a coal which upon combustion yields
2.0σ. Natural gas produces negligible SO_2. To produce one million BTUs
from gas and coal, yet maintain 1.5 lbs SO_2 simply requires using a 25
to 75 percent ratio of gas energy release to coal energy release. To
calculate the SO_2 emissions of the coal or oil we simply divide 10^6 BTU
by the combustion energy/lb and multiply the result by 2S, where S is
the fractional weight of sulfur in the coal. The factor of 2 takes into
account the fact that SO_2 is twice the weight of sulfur. Table 4.2
gives the percentage of gas that must be burned in conjunction with coal
of categories A, B, C, and D to meet various possible emissions
standards (SIPs) listed in the first column.

Table 4.2 Percentage of Natural Gas Needed for Various
Grades of Coal to Meet Possible SIP's (σ = lbs/10^6BTU)
*(denotes best available control technology BACT)

SIP	A(1.2σ)	B(2.4σ)	C(4.2σ)	D(6.0σ)
0.6	50	75	86	90
0.8	33	67	81	87
1.0	17	58	76	83
1.2*	0	50	71	80
1.4	0	42	67	77
1.6	0	33	62	73
2.0	0	17	52	67
2.4	0	0	43	60
2.8	0	0	33	53

The low ash content of oil (typically 0.1 percent) is one feature of oil burning that is not easily achieved by burning a mixture of gas (which has no ash) and coal (which has ash ranging from 5 to 25 percent. It is not possible simply by a choice of reasonable proportions to reduce the ash production of a gas-coal mixture to the ash production of oil. Fortunately, however, precipitators capture fly ash in the flue gas with very high efficiency (99 percent) at a relatively low cost. Furthermore, techniques such as physical or chemical coal cleaning or the use of cyclone furnace injectors can be used to minimize the ash injected into the boiler. The most difficult aspect of replacing oil with gas-coal can thus be accommodated if the gas-coal burning system is designed together with ash suppression systems.

For the purposes of economic analyses we will choose the following representative scenario for retrofitting an existing oil boiler. Experience to this point (Kirkwood et al. 1978) suggests that a conversion to coal alone of an average oil boiler would require a derating to about 60 percent of the nameplate oil boiler capacity. By burning gas and coal in the ratio of 25 to 75 percent by weight, it would appear possible to work the boiler up to 80 percent of the nameplate capacity. Present experience suggests that this would handle most of the annual use of the oil boiler. To achieve higher power levels up to nameplate capacity we step the gas proportion upward to 40 percent of rated capacity, while maintaining the coal power at 60 percent. Our scenario averaged over base load and peak periods corresponds approximately to a 70/30 ratio of coal to gas. Such a retrofit scenario would have a much greater reliance upon coal than the coal-oil mixtures (COM) approach, wherein oil, the premium fuel, supplies about 70 percent of the power and coal 30 percent. This scenario illustrates the fundamental aspect of our gas-coal approach, namely that the gas mainly serves to facilitate the use of coal, the primary fuel.

Because of the intermediate characteristics of oil, it is reasonable to expect in all matters with the exception of ash problems that a gas-coal mixture fed into the burning zone of an oil boiler would burn somewhat like oil. The fly ash and particulate emissions problem can be handled by the use of highly efficient precipitators or bag houses. However ash problems such as slagging, fouling, increased corrosion, tube erosion, plugging of air passes require more complex measures (Babcock and Wilcox 1978, Combustion Engineering 1981). Means of dealing with the ash problems are just being developed as various approaches to oil boiler retrofitting are now being examined, tested or carried out. These approaches include (1) conversion of coal-oil mixtures such as recently undertaken at the Sanford Station in Florida (Cook, 1980) (2) conversion to coal use with the maintenance of oil firing capability as recently carried out at the Kwinana power station in Perth, Australia (Kirkwood et al. 1978), (3) the installation of a new coal boiler to be used with the remainder of the electric plant (Philipps 1979, 1980; Ehrlich, Drenker and Manfred 1980) and (4) the use of coal-water slurries (ibid., Glenn 1979). In addition various types of clean fuels derived from coal using hydroliquefaction processes or chemical coal cleaning or coal gasification have been proposed for use in oil boilers. However for the most part these will not be available until the 1990s. The gas-coal approach is the most recently proposed approach to oil-backout and a paucity of literature exists on it. From a broad perspective the gas use is intended to facilitate the clean burning of coal which is the primary fuel.

Gas-coal use to replace oil may be viewed as a conservation measure in several important ways. First, by minimizing conversion capital costs while backing out of expensive oil we conserve both capital and consumer resources. Second, by extending the useful life of existing oil plants, say for 15 years, we make a lesser long term commitment to coal burning than does the construction of a new coal-fired plant.

An abbreviated commitment of capital to coal burning could be important if the problem of climate change by carbon dioxide emissions should develop at a faster pace than is generally anticipated. Over the last century carbon dioxide concentrations have increased about 18 percent or on the average of 0.2 percent per year, however, in the past decade the rate of increase has accelerated to about 0.3 percent per year. While a possibility exists that the increase may have other causes, such as deforestation, the works on this subject (National Academy of Sciences 1977, 1979; Bach 1980, Kellogg 1978, 1981) link the increase in part to increased combustion of fossil fuels Furthermore most climate change experts agree that a doubling of atmospheric CO_2 would lead to global temperature increases of the order of 3°C. Coastal geographic states, such as Florida, would be particularly vulnerable to ice cap melting associated with such an increase. The replacement of oil by coal power alone would further exacerbate the potential climate change problem since the combustion products would have more CO_2 and less H_2O. On the other hand, by burning mixtures of natural gas and coal we could approximately maintain the proportions of CO_2 and H_2O resulting from oil combustion. Granting the speculative nature of climate change theories the gas-coal option would, in effect, leave things unchanged for two decades or so while we develop a clearer understanding of possible adverse effects of fossil fuel burning and a clearer overall long-term energy strategy.

Table 1.5 summarized the various properties of gas, oil and coal which support the thesis that gas and coal can be burned simultaneously to displace oil.

In the next chapter we will examine the costs of various oil back-out options.

CHAPTER 5

CONVERSION COSTS OF OIL BACK-OUT OPTIONS

The first section of this chapter classifies all of the state's
utility oil boilers as to their convertibility to coal burning. The
second section presents itemized estimates of conversion costs of
various oil backout options for a 200 megawatt plant based upon the
methodology and economic estimates of Burns and Roe studies (Philipp
1980, Philipp and Kregg 1980). The third section contains independent
conversion cost estimate of various coal backout options using component
costs data provided by an I.C.F. Inc. Report (1980). The final section
gives a simplified economic analyses for the various conversions of a
200 megawatt plant. The results provide a relative financial assessment
of various oil backout options.

5.1 A Classification of Florida's Utility Oil Boilers

To facilitate the estimation of costs it is helpful first to
develop a scale for the convertibility of oil boilers. As a concrete
example, we have categorized Florida municipal and investor-owned power
company boilers now used to produce steam to drive steam turbines for
electricity generation into five major groups, as itemized in Table
5.1. (FCG 1980, Ten-Year Site Plans 1981, DOE 1978, 1979). Each
category selection is based on the following criteria:
Group A: Oil boilers located where existing or projected coal units
are available. Because coal and ash handling are already accommodated,
these sites are logical candidates for the earliest conversions from oil
to coal.
Group B: Oil boilers in service approximately 25 years or less with
adequate site size and/or port facilities to accommodate coal and ash
handling.

Table 5.1 Florida's Oil Fired Steam Boilers for Electric Power Generation

Category A: Highly Suitable, Coal Facilities Already Available
(Total MCR = 2720 MW)

Utility Co.	Plant Location	Unit	Name Plate (MW)	Fuel Pri	Fuel Alt*	Service Date	Net Cap Sum	Net Cap Win	Flue Gas Cleaning SOx	Part	NOx	Fuel Transport Pri.	Alt.	Land Area Total	Inuse	Present or Projected Coal Units MCR(MW)
Gainesville RUB	Deerhaven Alachua Co.	{ 1	75	HO	NG	8/72	82	84	-	-	-	TK	PL	1116	105	1 @ 235
FPL	Sanford Volusia Co.	3	156	HO	NG	5/59	137	139	CSCF	CSCF	-	B	PL	1647	1607	1 @ 366
		4	436	HO	-	7/72	362	366	CSCF	MC	-	B	-			
		5	436	HO	-	6/73	362	366	CSCF	MC	-	B	-			
	Martin Martin Co.	1	863	HO	10	12/80	783	790	CSCF	MC	FGR	PL	-	11,242	9520	2 @ 710
		2	863	HO	-	OnLine 1981	783	790	CSCF	MC	FGR	PL	-			
Lakeland Dept. Elec & Water	McIntosh Polk Co.	1	104	HO	NG	1970	92	93	HS	-	-	TK	PL	275	25	1 @ 334
		2	126	HO	-	1976	104	102	LS	-	FGR	TK	PL			

Codes: HO – Heavy Oil; NG – Natural Gas; CSCF – Controlled Sulfur Content of Fuel; MC – Mechanical Collector; FGR – Flue Gas Recirculation; B – Barge; PL – Pipeline; TK – Truck; WA – Water Borne; LS – Low Sulfur; HS – High Sulfur (within maximum limits).

*A number denotes maximum distance to natural gas pipeline for plant as a whole.

MCR denotes Maximum Continuous Rating, an average of summer and winter net capacity.

Table 5.1 cont'd

Category B: Suitable for Conversion

Category B1: (100 Acres or Greater) (Total MCR = 6776 MW)

Utility Co.	Plant Location	Unit No.	Max. NamePlate Capacity (MW)	Fuel PRI	Fuel ALT	Service Date	Net Cap. (MW) Summer	Net Cap. (MW) Winter	Flue Gas Cleaning SOx	Flue Gas Cleaning Part	Flue Gas Cleaning NOx	Fuel Transport PRI	Fuel Transport ALT	Land Area Total	Land Area In Use
FPL	Fort Myers Lee County	2 / 1	402 / 156	HO / HO	50 / –	7/69 / 12/58	367 / 137.0	370 / 139.0	CSCF / CSCF	MC / CSCF	– / –	B / B	– / –	460	460
	Turkey Pt. Dade County	1 / 2	402 / 402	HO / HO	NG / NG	4/67 / 4/68	367 / 367	370 / 370	CSCF / CSCF	MC / MC	– / –	B / B	PL / PL	11,967	11,967
	Manatee Manatee County	1 / 2	863 / 863	HO / HO	10 / –	10/76 / 12/77	783 / 783	790 / 790	CSCF / CSCF	MC / MC	FGR / FGR	PL / PL	– / –	11,702	7,747
FPC	Bartow Pinellas County	1 / 2 / 3	128 / 128 / 239	HO / HO / HO	– / – / NG	9/58 / 8/61 / 7/63	108 / 117 / 205	109 / 119 / 209	– / – / –	– / – / –	– / – / –	WA / WA / WA	– / – / PL	1,348	1,245
	Anclote Pasco County	1 / 2	556 / 556	HO / HO	10 / –	11/74 / 10/78	511 / 511	517 / 517	– / –	– / –	– / –	PL / PL	– / –	405	376
Jacksonville Electric Authority	Northside Duval County	1 / 2 / 3	298 / 298 / 564	HO / HO / HO	20 / – / –	11/66 / 3/72 / 7/77	262 / 262 / 499	262 / 262 / 499	LS / LS / LS	– / – / –	BM / – / BM	WA / WA / WA	– / – / –	493	204
Orlando Utilities Commission	Indian River Brevard County	1 / 2 / 3	87 / 208 / 344	HO / HO / HO	NG / NG / NG	2/60 / 12/64 / 2/74	88 / 204 / 318	90 / 208 / 321	LS / LS / LS	– / – / –	– / – / –	WA / WA / WA	– / – / –	100	20
City of Tallahassee	Hopkins Leon County	1 / 2	75 / 259	HO / HO	NG / NG	5/71 / 10/77	55 / 220	77 / 220	LS / LS	– / –	– / BM	TK / TK	PL / PL	231	35

Table 5.1 cont'd

Category B1: (100 Acres or Greater) - CONTINUED

Utility Co.	Plant Location	Unit No.	Max. Name Plate Capacity (MW)	Fuel PRI	Fuel ALT	Service Date	Net Cap. (MW) Summer	Net Cap. (MW) Winter	Flue Gas Cleaning SO_x	Part	NO_x	Fuel Transport PRI	Fuel Transport ALT	Land Use Total	Land Use In Use
FPL	Lauderdale Broward County	4	156	HO	NG	9/57	137	139	CSCF	CSCF	-	PL	PL	764	736
		5	156	HO	NG	4/58	137	139	CSCF	CSCF	-	PL	PL		
	Palatka Putnam County	2	75	HO	NG	8/56	75	75	CSCF	CSCF	-	B	PL	126	126
FPC	Turner Volusia County	3	79	HO	NG	11/59	70	72	-	-	-	TK/WA	PL	127	122
		4	82	HO	NG	5/59	71	73	-	-	-	TK/WA	PL		
	Suwannee Suwannee County	3	75	HO	NG	10/56	80	80	-	-	-	TK	PL	596	596

Category B2: (Small Sites with Port - Greater Than 150 MW) (Total MCR-2560 MW)

Utility Co.	Plant Location	Unit No.	Max. Name Plate Capacity (MW)	Fuel PRI	Fuel ALT	Service Date	Net Cap. (MW) Summer	Net Cap. (MW) Winter	Flue Gas Cleaning SO_x	Part	NO_x	Fuel Transport PRI	Fuel Transport ALT	Land Use Total	Land Use In Use
FPL	Cape Canaveral Brevard County	1	402	HO	NG	4/65	367	370	CSCF	MC	-	B	PL	98	97
		2	402	HO	NG	5/69	362	366	CSCF	MC	-	B	PL		
	Port Everglades Broward County	1	225	HO	NG	6/60	204	206	CSCF	MC	-	Ship	PL	92	92
		2	225	HO	NG	4/61	204	206	CSCF	MC	-	Ship	PL		
		3	402	HO	NG	7/64	367	370	CSCF	MC	-	Ship	PL		
		4	402	HO	NG	4/65	367	370	CSCF	MC	-	Ship	PL		
	Riviera Palm Beach County	3	310	HO	NG	6/62	272	275	CSCF	MC	-	Ship	PL	22	22
		4	310	HO	NG	3/63	272	275	CSCF	MC	-	Ship	PL		
Jacksonville Electric Authority	Kennedy Duval County	10	150	HO	10	12/61	130	136	LS	-	-	WA	-	53	26

Table 3.1 cont'd

Category B3: (Small Site With Port - Less than 150 MW) (Total MCR - 376 MW)

Utility Co.	Plant Location	Unit No.	Max. NamePlate Capacity (MW)	Fuel PRI	Fuel ALT	Service Date	Net Cap. (MW) Summer	Net Cap. (MW) Winter	Flue Gas Cleaning SOx	Part	NOx	Fuel Transp. PRI	Fuel Transp. ALT	Land Use Total	Land Use In Use
Key West Utility Board	Key West Monroe County	5	20	HO	90	1966						WA	-		
	Stock Island Monroe County	1	37	HO	90	1972						WA	-		
City of Tallahassee	Purdom Wakulla County	5 6 7	25 25 50	HO HO HO	NG NG NG	4/58 1961 1966	22 22 45	22 22 46	- - -	- - -	- - -	WA WA WA	PL PL PL	42	38
Fort Pierce Utilities Auth.	King St. Lucie Cty.	7 8	37 56	HO HO	- NG	1964 1976	35 51	35 51				WA WA	- PL		
Jacksonville Electric Authority	Kennedy Duval County	9	50	HO	10	1/58	43	46	LS	-	-	WA	-	53	26
	Southside Duval County	4	75	HO	10	11/58	70	71	LS	-	-	WA	-	38	24

Category C: Small Site (Total MCR = 247 MW)

Utility Co.	Plant Location	Unit No.	Max. Name Plate Capacity (MW)	Fuel PRI	Fuel ALT	Service Date	Net Cap. (MW) Summer	Net Cap. (MW) Winter	Flue Gas Cleaning SOx	Part	NOx	Fuel Transp. PRI	Fuel Transp. ALT	Land Area Total	Land Area In Use
Lakeland Dept. Elec. & Water	Larsen Polk County	7	50	HO	NG	1966	49	52	HS	-	-	TK	PL	9	9
Gainesville RUB	Kelly Gainesville Alachua County	7 8	25 50	HO HO	NG NG	8/61 4/65	22 51	22 51	- -	- -	- -	TK TK	PL PL	11	11
Vero Beach Municipal	Vero Beach Indian River County	1 2 3 4	13 20 33 56	HO HO HO HO	10 - - -	1961 1964 1971 1976	13 17 33 56	13 17 33 56							

Table 5.1 cont'd

Category D: Remaining Oil Boiler (CONTINUED)

Utility Co.	Plant Location	Unit No.	Max. Name Plate Capacity (MW)	Fuel PRI	Fuel ALT	Service Date	Net Cap. (MW) Summer	Net Cap. (MW) Winter	Flue Gas Cleaning SOx	Part.	NOx	Fuel Transp. PRI	Fuel Transp. ALT	Land Area Total	Land Area In Use
Util. Brd. of Key West	Key West Monroe County	1	6	Oil	–	1953								9	9
		2	6	Oil	–	1953									
		3	19	Oil	–	1957									
		4	19	Oil	–	1957									
Lakeland Dept. Elec. & Water Util.	Larsen Memorial Polk County	4	20	HO	NG	1/50	19	20	HS	–	–	TK	PL	9	
		5	25	HO	NG	1/56	24	24	HS	–	–	TK	PL		
		6	25	HO	NG	1/59	24	25	HS	–	–	TK	PL		
Orlando Util. Commission	Lake Highland Orange County	1	29	HO	NG	9/49	30	31	LS	–	–	TK	PL	10	7
		2	38	HO	NG	9/54	30	31	LS	–	–	TK	PL		
		3	38	HO	NG	6/56	30	31	LS	–	–	TK	PL		
City of Tallahassee	Sam O. Purdom Wakulla County	1	8	HO	–	1/52	0.0*	0.0	LS	–	–	WA	–	42	38
		2	8	HO	–	1/52	0.0*	0.0	LS	–	–	WA	–		
		3	8	HO	NG	1/52	7	8	LS	–	–	WA	PL		
		4	8	HO	NG	4/54	7	8	LS	–	–	WA	PL		
Tampa Elec. Co.	Hookerspoint Hillsborough County	1	33	HO	–	7/48	24	24	LS	–	–	WA	–	25	25
		2	34	HO	–	6/50	24	24	LS	–	–	WA	–		
		3	34	HO	–	8/50	24	24	LS	–	–	WA	–		
		4	49	HO	–	10/53	38	38	LS	–	–	WA	–		
		5	82	HO	–	5/55	67	67	LS	–	–	WA	–		

* Cold Standby

Table 5.1 cont'd

Category D: Remaining Oil Boilers (Total MCR - 1158 MW)

Utility Co.	Plant Location	Unit No.	Max. Name Plate Capacity (MW)	Fuel PRI	Fuel ALT	Service Date	Net Cap. (MW) Summer	Net Cap. (MW) Winter	Flue Gas Cleaning SO$_x$	Part.	NO$_x$	Fuel Transp. PRI	Fuel Transp. ALT	Land Area Total	Land Area In Use
FPL	Cutler Dade County	5	75	HO	NG	11/54	67	70	CSCF	MC	–	B	PL	96	96
		6	162	HO	NG	7/55	130	132	CSCF	MC	–	B	PL		
	Palatka Putnam County	1	34	HO	NG	8/51	32	34	CSCF	CSCF	–	B	PL	126	126
	Riviera Palm Beach Co.	1	44	HO	NG	11/46	40	41	CSCF	MC	–	S	PL	22	22
		2	75	HO	NG	11/53	69	71	CSCF	MC	–	S	PL		
FPC	Avon Park Highlands Co.	2	46	HO	NG	11/52	38	38	–	–	–	TK	PL	56	40
	Higgins Pinellas Co.	1	46	HO	NG	6/51	40	42	–	–	–	WA	PL	142	79
		2	46	HO	NG	6/53	39	41	–	–	–	WA	PL		
		3	46	HO	–	1/54	41	43	–	–	–	WA	–		
	Suwannee Suwannee Co.	1	40	HO	NG	11/53	33	34	–	–	–	TK	PL	596	596
		2	38	HO	NG	11/54	32	33	–	–	–	TK	PL		
Ft. Pierce Util. Auth.	King St. Lucie Co.	5	8	HO	NG	1/53	7	7	–	–	–				
		6	18	HO	–	1/58	18	18	–	–	–				
Gainesville RUB	Kelly J.R. Alachua County	5	11	HO	NG	1955	14	14	–	–	–	TK	PL	11	11
		6	14	HO	NG	3/58			–	–	–	TK	PL		
Jacksonville Elec. Auth.	Kennedy Duval County	8	50	HO	–	7/55	43	46	LS	–	–	WA	–	53	26
	Southside Duval County	1	38	HO	–	11/50	26	27	LS	–	–	WA	–	38	24
		2	38	HO	–	1/51	26	32	LS	–	–	WA	–		
		3	50	HO	–	1/55	46	48	LS	–	–	WA	–		

Group B_1: Sites of 100 acres or greater.

Group B_2: Sites of less than 100 acres (small sites) but with port facilities. Oil boiler capacity of 150 megawatts or greater.

Group B_3: Small site with port facilities and oil boiler capacity of less than 150 megawatts.

Group C: Oil boilers in service approximately 25 years or less that are located on small sites with no port facilities available. All were found to have boiler capacity of less than 150 megawatts.

Group D: All remaining Florida steam turbined oil boilers that have been in service over 25 years. Due to life expectancy these boilers have not been priced for conversion.

Group E: All remaining gas turbine generators that primarily use distillate oil; these are not listed, but they have a maximum total nameplate capacity of 4,889 megawatts.

5.2 Conversion Costs

Table 5.2 presents cost of converting an oil-fired boiler to various coal-mixture fuels. The cost figures are both itemized and totalled; the itemized cost figures were normalized into dollars per kilowatt net so that the costs of various fuel alternatives could be compared on an accurate relative basis. The cost figures used in this section were drawn and modified from the itemized costs of a new coal-fired boiler, which were obtained from Burns and Roe (Philipp and Kregg 1980, Philipp 1980). The columns headed "COM" (coal-oil mixture) and "Dual Coal/Oil" also were taken, without modification, from the Burns and Roe report. The costs in the remaining columns are projected from the new coal-fired boiler costs.

For four additional options considered, boiler conversion costs are estimated at $47 per kilowatt. This is a saving compared to the $115 per kilowatt figure for a new coal boiler. It allows only for boiler modification, e.g., new tube banks, blowers, and ash removal and pulverizer equipment. The modified boiler cost is identical to the $47 per kilowatt for the dual coal/oil conversion.

Table 5.2 Conversion Cost Estimates of Various Coal Backout Options in $/KW for a 200 MW Plant

Capital Cost Items	COM	Dual Coal/Oil	New Coal	Gas/Coal A	Gas/Coal B	Gas/Coal C	Coal/Water**
1. Coal Handling & Storage	11+	18	20	5	20	20	11
2. Excavations & Foundations	7	10	12	5	12	12	12
3. Ash Handing System	10	12	14	5	14	14	14
4. Electrostatic Precipitator	22	25	40	40	40	40	40
5. Miscellaneous	6	8	12	12	12	50*	20
Sub-Total	56	73	98	67	98	136	97
6. Owner's Cost/20%	11	15	20	13	20	27	19
Sub-Total	67	88	118	80	118	163	116
Sub-Total @ 60% Coal	--	--	--	48	71	98	--
Sub-Total @ 65% Coal	--	--	--	--	--	--	76
7. Boiler Conversion	14	47	115	47	47	47	47
8. Owner's Cost/20%	3	9	23	9	9	9	9
TOTAL COST (without F.G.D.)	84	144	256	104	127	154	132
9. F.G.D.	--	72	120	--	--	--	--
TOTAL COST (with F.G.D.)	--	216	376	--	--	--	--

* 38 $/KW for additional train or truck transportation equipment
+ It is assumed that COM fuel is purchased rather than prepared on site
** The cost of purchasing power to replace the derated power is not included

For plants that already have one or more of their boilers coal fired, coal handling and storage costs are discounted to $5 per kilowatt. Such plants already have the major components of the coal handling equipment available for use. Power plants that do not have existing coal handling facilities are assigned the same $20 per kilowatt of a new coal plant. A coal-water facility has been assigned the $11 per kilowatt figure for a coal-oil mixture plant. This figure includes the cost of laying the pipeline and coal handling and storage cost.

The costs of excavations and foundations are estimated at $5 per kilowatt for plants that already have one or more of their boilers coal fired and thus already have the major foundations and excavations for coal handling completed. All plants that have no existing coal handling equipment on site and thus require all of the necessary excavations and foundations are assigned the $12 per kilowatt of a new coal plant.

The costs of an ash handling system are taken as $5 per kilowatt for plants that already have a large coal fired boiler on site. Plants that do not have ash handling facilities and thus require the purchase of all necessary facilities are assigned the $14 per kilowatt of a new coal plant.

The costs of an electrostatic precipitator (EP) are $40 per kilowatt taken from the new coal plant cost since additional ash per kilowatt should be the same in all cases.

The miscellaneous cost now is a catch-all that includes installation and delivery charges and any other costs that might be incurred through errors or accidents. Class A and B gas-coal plants are assigned $12 per kilowatt, the new coal plant estimate. The $50 per kilowatt figure applies to plants that do not have coal or ash handling equipment or the on-site land for these facilities. Therefore, this figure also includes the price of land, rail spurs, rail cars, and locomotive for off-site coal and ash storage. The coal-water mixture approach is assigned a $20 per kilowatt figure. This figure includes some of the additional capital costs for pulverizing coal to an ultrafine consistency.

Owner's costs, which include finance charges, are estimated at 20 percent of the subtotal capital costs. Flue gas desulfurization costs

are not applied to the gas-coal and coal-water proposed conversions
since a low-sulfur coal is chosen.

As mentioned earlier, the cost figures used are projections from a
new 200 megawatt coal-fired boiler as estimated by Burns and Roe. When
we apply these figures to conversion we must consider the power derating
problem. A power plant converted from oil to coal burning experiences
a power derating to about 60 percent of its oil-fired design output
(Philipp 1980, Philipp & Kregg 1980, Power Systems Services 1980,
Kirkwood et al. 1978, Cantrell 1980). One cause of this derating is the
lower calorific value of coal ($\sim$ 12,000 BTU/lb) than oil ($\sim$ 18,000
BTU/lb) as indicated in Table 1.5. A second cause is the slag build up
on heat exchanger walls which infuences heat transfer rates to the steam
tubes. Thus a coal fired boiler is typically larger than an oil fired
boiler of the same power capacity as illustrated in Fig. 4.1. Therefore
upon converting an oil boiler to burning eastern bituminous coal one
should anticipate a derating which would permit burning only about 60%
of the coal required in a coal boiler of the same nameplate rating as
the original oil boiler. The matter is not cut and dried, however,
since for example, a generously sized oil boiler might do better.
Furthermore the flow field arrangement and the ash removal arrangement
will influence matters.

For specificity in this work we will use the Kwinana derating
factor of 0.6. Thus the subtotal of all directly related coal features
of the converted oil boiler (items 1 through 6 in Table 5.2) are multiplied by 0.6. To be conservative the boiler conversion costs, however,
have been taken at $47/KW as used by Burns and Roe which probably re-
flects the Kwinana experience.

The coal-water fuel which is currently under consideration is a
very fine pulverized coal in a ratio of 60-70 percent coal and 40-30
percent water by weight (Baumeister et al. 1978, Maize 1981). It is
directly injected, without dewatering into the combustion zone. Thus,
the coal handling, pulverizing, and burning equipment will be similar to
oil firing equipment. It is assumed that the coal water mixture is
prepared off the immediate boiler site and is pumped through a pipe to
the boiler. By burning a coal-water fuel mixture we encounter the
following questions:

1. How much extra energy is consumed in pulverizing the coal to say 90 percent through 320 mesh?
2. How much additional energy is required to vaporize and heat the inert water, used in transporting the pumpable mixture from the pulverizer-mixer to the high temperature furnace?
3. How much derating will result from the lowered equivalent calorific value of the coal/water mixture?
4. How is the derating influence by the higher quality physically cleaned coal which should be used for coal/water mixtures?

The answers to these questions are not yet clear since experimentation on coal/water mixtures is just underway. However it would appear that:

1. The energy consumed in pulverizing coal to 90 percent through a 320 mesh would be about 4-6 percent of the plant's gross output as compared to about 1 percent for the conventional 90 percent through 200 mesh (Ehrlich et al. 1980).
2. The loss of energy due to the evaporation and heating of water is about 4-5 percent of the plant's gross output.
3. The derating expected is significantly influenced by the calorific content. If we mix a high grade (14,000 BTU/lb) coal at 1.3 relative density with 30 percent by weight of water at a density of one, we in effect produce a fuel of about 8,000 BTU/lb calorific value, similar to lignite.
4. Because of the higher quality coal which has some ash removed, the influence of ash upon the derating will be reduced. Furthermore the smaller particle size would reduce the collisions with the heater banks and reduce the slagging.

Considering all of these factors we guesstimate a derating to about 65 percent of the original nameplate capacity. However if the original oil capacity was assigned conservatively (i.e. the boiler was generously sized), it might be possible to achieve a higher percentage of nameplate capacity.

In our economic analysis the energy loss due to vaporization is included in the fuel cost and is simply the standard $ per ton plus an extra 8 percent to account for the difference in energy consumption

experienced due to higher pulverizing and heating and evaporation
losses. Derating should be quantified by incorporating the overall
costs of a peaking unit to replace the lost capacity. On the other hand
if the derating is to the 65-75 percent range and the use of the
generator is mostly below 60 percent, the loss of capacity might be
accommodated by purchases from other units of an interconnecting grid if
a surplus capacity is available.

For calculations on the coal-water mixture we used a derating to 65
percent of the oil fired capacity. We therefore multiply the total cost
by 0.65 to get the final dollar per kilowatt conversion cost for the
coal conversion features of the coal-water approach. The boiler costs
are not multiplied by a derating factor to avoid overestimating the
conversion savings associated with derating.

The total cost values are the total conversion costs per kilowatt
and are to be multiplied by a plant's specific nameplate kilowatt rating
to get the specific conversion cost. These values are based strictly on
coal-related costs. They do not consider gas hook-up costs. Also,
these conversion costs remain constant for any gas/coal fuel ratio.

The factor would suggest that the coal-water mixture approach is
relatively inexpensive. It must be noted, however, that 35 percent of
the original oil burning capacity would be sacrificed unless the oil
burning capability is retained when load demand exceed 65 percent or
surplus capacity is available in the grid. For the gas-coal water
approach, the high calorific value of the natural gas permits packing
more power into the original oil boiler so that little derating might be
expected. The gas-coal-water approach should thus retain the
pumpability benefits of the coal-water system while avoiding the large
replacement costs of the lost capacity associated with the strict coal-
water approach.

5.3 Independent Conversion Cost Estimate

We have carried out an independent conversion cost estimate of
various coal backout options for a 200 megawatt plant so that our study
is not entirely dependent upon cost data published by Burns and Roe,
Inc. (Philipp 1980). Some of the cost data necessary for this

verification are provided in a report written by I.C.F. Inc. (1980) which appears to involve independent cost estimates.

The I.C.F. report presented the itemized components, and their costs, required in the construction of a new coal fired boiler as prescribed in Figure 5.1, from the point of fuel entry onto the site to the point of steam exit, exclusive of fuel and fuel transportation (BW 1978, CE 1981). These cost data were then extrapolated to a 200 megawatt plant using various scale factors. The scaling equation is as follows:

$$\frac{I_1}{I_2} = [\frac{C_1}{C_2}]^p \qquad\qquad 5.1$$

where I_1 = the unknown total investment conversion cost at a new megawatt rating, C_1 = the new megawatt rating, I_2 = the total investment at a specific megawatt rating, C_2 = the specific megawatt rating (200 megawatt in this case), and p = scale factor (0.8 in this case). The extrapolated values were then modified to accommodate the natural gas-coal conversions. These values are presented in Table 5.3.

It is gratifying to see that all of the columns, except the Natural Gas-Coal water mixture, were within 5 percent of our projections based upon the Burns and Roe cost figures. The coal/water mixtures figures were within 15 percent. In view of this verification of our earlier projection we will continue to use Table 5.2 in our economic analysis.

5.4 Simplified Economic Analyses of Oil Back-Out Options

To obtain an initial assessment of the various conversion options for oil boilers we first have carried out calculations that parallel a study by Philipp (1980). Table 5.4 is an economic comparison of seven alternatives for steam turbine oil boilers for a 200 megawatt unit. From left to right, these alternatives are: coal-oil mixture, dual use of coal and oil, coal, the A, B, and C categories of coal with select use of gas, coal/water, and finally, gas-coal-water. The table is set up in the form of a financial statement, with four major parts for consideration: (1) conversion costs for each alternative, taking into account the percentages use of relevant fuels; (2) annual costs,

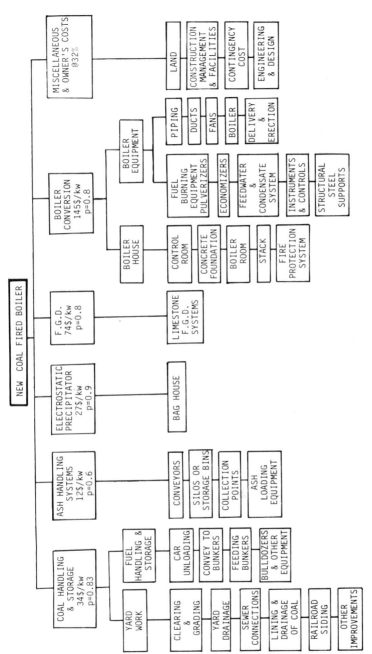

Fig. 5.1 Components Required in the Construction of a New Coal Fired Power Plant.

Table 5.3 Independent Conversion Cost Estimates of Various Coal Backout Options in $/KW for a 200 MW Plant

Capital Cost Items	COM	Dual Coal/Oil	New Coal	Gas/Coal A	B	C	Coal/Water**
1. Coal Handling & Storage	11	18	34	8	34	54*	11
2. Excavations & Foundations	7	10	12	5	12	12	12
3. Ash Handling System	10	12	12	5	12	12	12
4. Electrostatic Precipitator	22	25	27	27	27	27	27
Sub-Total	50	65	85	45	85	105	62
Sub-Total @ 60% Coal	--	--	--	27	51	63	--
Sub-Total @ 65% Coal	--	--	--	--	--	--	40
5. Boiler Conversion	14	47	133	50	50	50	50
Sub-Total 2	64	112	218	77	101	113	90
6. Misc. & Owner's Cost @32% for Sub-Total 2	20	36	70	25	32	36	29
TOTAL COST (without F.G.D.)	84	148	288	102	133	149	119
7. F.G.D.	--	72	74	--	--	--	--
8. Misc. & Owner's Cost (@32%) for F.G.D.	--	--	24	--	--	--	--
TOTAL COST (with F.G.D.)	--	220	386	--	--	--	--

* 20 $/KW for additional train or truck transportation equipment
** The cost of purchasing power to replace the derated power is not included

Table 5.4 Economic Comparisons of Alternatives (200 MW Unit)**

Economic Considerations	Oil	COM.	Dual NFGD	Coal/ Oil FGD	New Coal FGD	Gas/ Coal (A)	Gas/ Coal (B)	Gas/ Coal (C)	Coal/ Water*	Gas/ Coal/ Water
Annual Coal Consumption, % by Energy	0	30	80	80	100	70	70	70	100	70
Conversion Cost										
Direct Cost, $/KW	--	84	144	216	376	104	127	154	132	132
Direct Cost, $10^6$$	--	16.8	28.8	43.2	75.2	20.8	25.4	30.8	26.4	26.4
Direct Cost+I DC, $10^6$$	--	20.3	34.8	52.3	91.0	25.0	30.5	37.0	31.7	31.7
Annual Cost										
Fuel Costs	46.0	39.4	22.0	22.9	16.6	19.2	19.2	19.2	23.8	21.9
O & M Costs	Base	0.4	0.5	1.0	1.2	0.5	0.5	0.5	1.2	1.2
Total Operating Costs	46.0	39.8	22.5	23.9	17.8	19.7	19.7	19.7	25.0	23.1
Differential Operating Costs	Base	-6.2	-23.5	-22.1	-28.2	-26.3	-26.3	-26.3	-21.0	-22.9
Fixed Charges	--	4.06	6.96	10.46	18.2	5.0	6.1	7.4	6.3	6.3
Total Annual Costs	46.0	43.86	29.46	34.36	36.0	24.7	25.8	27.1	31.3	29.4
Differential Annual Cost	Base	-2.14	-16.54	-11.64	-10.0	-21.3	-20.2	-18.9	-14.7	-16.6
Payback Period	--	3.3	1.5	2.4	3.2	1.0	1.2	1.4	1.5	1.4

Notes: Annual Capacity Factor, 60%
       Fixed Charges, 20%/yr
       Interest During Construction, 10%/yr
       Capital Costs annualized for 15 yr period
       **All Costs in million dollars per year

1980 Fuel Costs: Fuel Oil, 4.80 $/$10^6$ BTU (30$/BBL)
               Coal, 1.67 $/$10^6$ BTU (40$/Ton)
               COM, 4.11 $/$10^6$ BTU
               Coal with water, 47$/Tons + 8%
               Gas 2.60/10^6CF

*Costs for boiling water off and finer coal grind included but not replacment costs for lost power.

(3) fixed charges; and (4) payback period (DOE/OFC 1980, Resources for the Future 1980, AGA 1981).

The first of the parts is a summary of the direct and indirect costs associated with converting a steam turbine oil boiler. The gas/coal (B) column will be used as an example for interpretation; 127 is the dollar cost per kilowatt estimated for conversion with 70 percent coal, 30 percent gas use in an oil boiler. The sources of the values used for these direct costs have been explained in connection with Table 5.2. Multiplying $127 per kilowatt by 200,000 kilowatts generates $25.4 million (all subsequent costs are expressed in millions of dollars). The indirect costs (IDC) is generated by multiplying the direct costs by 20 percent to allow for overhead type costs.

The second part is basically self-explanatory. 19.2 is the fuel cost of 70 percent coal and 30 percent gas use in a 200 megawatt boiler. 0.5 is the annual operation and maintenance costs. 19.7 is the total annual operating costs. Operating costs are directly variable to the amount of output. Therefore, by subtracting the total operating cost from the base value 46 (oil column) the difference, 26.3, is the total differential operating savings.

The third part of this comparison is an estimate of fixed charges associated with each alternative. By adding this figure to the total annual variable costs, we generate the total annual cost figure of 25.8 for a category B, gas-coal burner. The differential annual cost incorporates fixed costs into the analysis. 20.2 is simply the subtraction of 6.1 (fixed charges) from 26.3 (differential operating costs) - all costs are before taxes.

The last part, payback period, following the Burns and Roe method, is simply the direct plus indirect costs divided by the differential operating costs (i.e. annual variable savings).

Since our concentration has been on the gas-coal option, we have not developed our categories A, B, and C for the other coal conversion options. Thus for comparative purposes the Category B gas-coal scenario should be used. Note that the differential annual costs for the gas-coal (B) is -20.2 which compares favorably with the other approaches, particularly the coal-oil approach.

CHAPTER 6

QUANTITATIVE INPUTS FOR ECONOMIC ANALYSES

The most important inputs for our economic analyses are the base
costs of fuel oil, coal and natural gas. Unless otherwise noted our
1980 base cost of fuel oil is taken at $30 per barrel and the 1980 base
cost of coal is taken at $40 per ton. These fuel costs are the same as
those used in a recent Burns and Roe report (Philipp and Kregg 1980) so
that utilizing these fuel costs facilitates intercomparison with their
alternate conversion cost options. A natural gas price of $2.60 per
million cubic feet for 1980 was selected as representative from
Department of Energy Reports on the Cost and Quality of Fuels for
Electric Utility Plants (1979, 1980, 1981) and the American Gas
Association TERA-II Model (AGA 1981).

To accommodate the time dependence of fuel costs and other important variables we have utilized analytical formulas to express extensive
economic tabular data in convenient and manageable forms. With readily
available micro and minicomputers, analytic formulas may be used to
reproduce data with arbitrary precision and to interpolate between data
points. In the following subchapters we propose simple analytical
formulas for important economic inputs. We adopted this approach to
avoid ambiguity and to set the stage for later sensitivity analyses.

6.1 Projected Fuel Price Analysis

Anticipated oil, gas, and coal prices have been obtained from the
output of projected fuel costs of the latest American Gas Association
TERA-II Model (AGA 1981). These values for the time interval from 1980
to 1995 are denoted by points on the semilogarithmic graph of Figure
1.5. The shapes of these curves appeared to be similar to an inverted
Verhulst distribution function (Green 1965, 1974), which may be

73

expressed by the formula

$$P_f = P_0 \left[\frac{e^{t/a} - \rho}{1 - \rho} \right]$$

6.1

where P_f = price of fuel ($\$/10^6 BTU$), T= year (1980-1995 interval) and
t = T-1980. The constants P_0, ρ and a have been determined by nonlinear
least square (NLLS). The resulting parameters for each fuel type are
given in Table 6.1.

Table 6.1   Fuel Price Parameters

	P_0	ρ	a	R	P_b
Coal	1.37	0.177	8.63	12.67	1.67
Oil	4.73	0.693	15.66	12.20	4.80
Gas	2.29	0.828	18.23	14.20	2.60
Inflation	1.00	0.305	15.44	8.08	

$$P_f = P_0 (e^{t/a} - \rho), \quad \bar{R} = 100 \ln(P_f/P_0)/\Delta t$$

The curve so represented by these parameters, are plotted as solid
lines in Figure 1.5. Also shown is an inflation function of the same
analytic form.

To determine the inflation function for 1980-1995 interval we first
determine the derivation of P_f with respect to T.

$$\frac{dP_f}{dT} = \frac{P_0 e^{t/a}}{a(1-\rho)}$$

6.2

Then the inflation rate (IR) is given by

$$IR = \frac{dP_f}{dT}/P_f = \frac{100}{P_f} \left[\frac{P_o e^{t/a}}{a(1-\rho)}\right] \qquad 6.3$$

which can be simplified to

$$IR = \frac{100}{a} \left(1 + \frac{\rho}{e^{t/a} - \rho}\right) \qquad 6.4$$

   Inflation rates parameters from the latest AGA TERA-II Model are
given in Table 6.2. The dates in parentheses denotes the midpoint used
in the inflation rate formula. Percentages in parentheses are values
obtained by our NLLS code with the adjusted parameters a and ρ in Table
6.1. These parameters together with $P_0 = 1.0$ fix the inflation function. This function is also shown in Figure 1.5.

Table 6.2   Inflation Rates

Tera (%)	Model(%)	Period (Annual)	Midpoint
8.7	(8.75)	from 1980 thru 1985	(1982.5)
8.1	(7.97)	from 1985 thru 1990	(1987.5)
7.4	(7.49)	from 1990 thru 1995	(1992.5)
7.1	(7.18)	from 1995 thru 2000	(1997.5)

The average growth rate $(\bar{R})$ can be calculated by

$$\bar{R} = \frac{1}{\Delta t} \int_0^{t_f} (IR)dt \qquad 6.5$$

Where t_f = final year growth considered, and Δt = difference between
final year and initial year. Performing the integration yields

$$\bar{R} = \frac{100}{\Delta t} \ln(e^{t/a} - \rho) \ \Big|_0^{t_f} \tag{6.6}$$

$$\bar{R} = \frac{100}{\Delta t} \ln \left(\frac{e^{t_f/a} - \rho}{1 - \rho}\right) = \frac{100}{\Delta t} \ln \left(\frac{P_f}{P_0}\right) \tag{6.7}$$

We evaluated the average growth rates for t_f = 15 (1980 to 1995) with Δt = 15 for coal, oil, gas and the inflation rate. These results are given in column headed R in Table 6.1. The last column headed P_b are the 1980 base prices of coal, oil, and gas equivalent to the $40/ton, $30/barrel, and $2.60/million cubic feet respective chosen as the base price in our subsequent calculations. These P_b values are in reasonable relationship to the P_0 values together with the time dependent parameters ρ and a for our actual calculations.

6.2 Dependence of Florida Coal Prices upon Sulfur Content

The average delivered prices of coal to Florida, by percent sulfur content for the years 1979, 1980, and 1981, were extracted from Department of Energy Reports on Cost and Quality of Fuels for Electric Utility Plants. Table 6.3 provides the price data obtained from this source with average coordinates selected in view of the percent sulfur interval spread. This data is plotted in Figure 6.1. A two-dimensional analytic formula was considered of the form:

$$P = \kappa e^{\alpha t} e^{-\beta x}, \tag{6.8}$$

where P is the price of coal in dollars per million BTU, t = T-1980 with T as the year and x is the percent sulfur content of coal. This non-linear equation can be linearized by taking its logarithm so that $\ln P = \ln \kappa + \alpha t - \beta x$. We used an available multiple linear regression program to fix the parameters κ = 2.13, α = 0.117 and β = 0.112. The plots of this function for the three years are shown as linear solid lines on the semilogarithmic graph in Figure 6.1. Note that κ = 2.13 denotes the price in 1980 for no-sulfur coal and that α = 0.117 can be considered as an inflation rate of 11.7 percent per year.

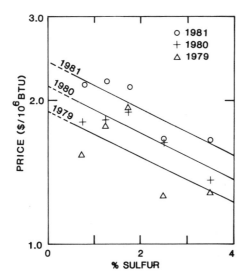

Figure 6.1   Dependence of Florida Coal Prices upon Sulfur Content and
Fit with Nonlinear Equation 6.8

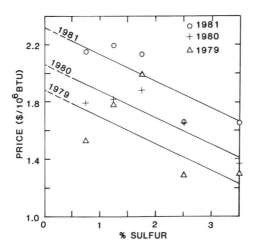

Figure 6.2   Dependence of Florida Coal Prices upon Sulfur Content and
Fit with Linear Equation 6.9

Table 6.3   Florida Coal Data

Percent Sulfur		Price (dollars/million BTU)		
Range	Data Point	1979	1980	1981
(>0.5 - 1.0)	0.75	1.52	1.79	2.15
(>1.0 - 1.5)	1.25	1.77	1.81	2.19
(>1.5 - 2.0)	1.75	1.94	1.88	2.13
(>2.0 - 3.0)	2.50	1.28	1.64	1.65
(>3.0)	3.50	1.29	1.37	1.65

Another multiple linear regression program was used following the more classical economic approach. Here we choose

$$P = a + bx + c_1 T_1 + c_2 T_2 + c_3 T_3 + \varepsilon \qquad 6.9$$

where P = Price of coal in dollars per million BTU, T_1 = 1 for 1979; 0 otherwise, T_2 = 1 for 1980; 0 otherwise, T_3 = 1 for 1981, 0 otherwise, x = percent sulfur content of coal, ε = error term.

The constants were determined to be a = -0.517, b = -0.186, c_1 = 2.40, c_2 = 2.53 and c_3 = 2.84 so that the price in one of the three years can be expressed on substitution as

$$P = -0.517 - 0.186x + 2.40T_1 + 2.58T_2 + 2.84T_3 \qquad 6.10$$

The R^2 statistic was .721. Figure 6.2 gives this representation.

6.3   Capital Conversion Costs Computation

To avoid the use of tables annual capital conversion costs (ACC) may be computed using

$$ACC = \frac{(CF)\ (TCC)}{CD} \qquad 6.11$$

where CF = compounding factor at i percent interest rate for n years, TCC = total capital conversion cost, CD = compounding factor for annual deposits at 1 percent interest rate for n years.

The expressions for CF and CD in terms of i and n are simply derived from compound and annuity formulas:

$$CF = (1 + i)^n \qquad\qquad 6.12$$

$$CD = \frac{(1+i)^n - 1}{i} = \frac{CF - 1}{i} \qquad\qquad 6.13$$

As an example, for an interest rate of 10 percent (i = .1) over a 15 year (n = 15) period, CF = 4.177 and CD = 31.772. Therefore, ACC = (.1315) (TCC).

6.4 Total Annualized Fuel Costs Computation

The total annualized fuel costs (TAFC) may be computed from annual fuel costs, as itemized in Table 1.4, by

$$TAFC = \frac{(1 + g)}{gn} [(1 + g)^n - 1] (L)(AC) \qquad\qquad 6.14$$

where g = net growth rate (NGR) above inflation rate, n = number of years, L = average load, and AC = annual fuel costs.

As an example, for oil at 4.1 percent NGR over 15 years with 1980 fuel costs of \$5.743 billion dollars then annualized fuel costs at a load factor of 60 percent would be:

$$TAFC = \frac{(1 + .041)}{(.041)(15)} [(1 + .041)^{15} - 1] (.6) (5.743 \times 10^9) \qquad\qquad 6.15$$

$$= 4.824 \text{ billion dollars}$$

6.5 Economy of Scale Correction

The simplified economic calculations made in Chapter 5 were made on the basis of estimated conversion costs rate (CCR) (in dollars per kilowatt). Thus, for application to the conversion of a given oil boiler this cost rate was multiplied by the nameplate power capacity. For more precise purposes we should first use a more realistic power rating such as the average maximum capacity rating (MCR) representing the means of the summer and winter MCRs. Secondly, we should allow for economies of scale since the many components of the conversion do not simply follow a fixed MCR. An equation for the total conversion cost (TCC) of the form

$$TCC = 200 \ (CCR) \times (MCR/200)^p \qquad\qquad 6.16$$

where the CCR has been estimated on the basis of a 200 megawatt unit is reasonable for most purposes. This can be rearranged in the more convenient form

$$TCC = (MCR) \ (CCR) \ [200/MCR]^{1-p} \qquad\qquad 6.17$$

where the quantity in the bracket now serves as the economy of scale factor. In our further calculations we choose $\bar{p} = 0.8$ based upon an average of all the scale factors for the component conversion costs of a new coal boiler.

CHAPTER 7

A DETAILED ECONOMIC ANALYSIS OF CONVERSION ALTERNATIVES

In comparing the merits of alternate investments, five factors are
of particular interest: the cost of capital, the discount rate, shutdown periods, future fuel prices, and the payback period. Each of the
factors influences the decision criteria of various conversion options
and requires certain economic assumptions.

7.1 The Cost of Capital
Several difficulties complicate assessment of cost of capital. As
examined by Miller and Modigliani (1976) in their classic study of the
electric utility industry, most of the difficulties remain unresolved.
The main problems involve measuring the influence of risk, uncertainty,
and opportunity costs; defining the value of an investment as a function
of appropriate variables, e.g. real rate of return on equity, debt
yield, taxes and/or replacement costs, and incorporating growth into the
analysis.

Most economic estimates of the cost of capital to a particular
plant focus on the rate of return necessary to allow the firm to (1) pay
a specified yield to its bondholders, (2) pay a satisfactory dividend to
its stockholders (i.e., return on equity), (3) cover the opportunity
costs of not using its capital for other purposes, and (4) account for
the impacts of inflation on 1, 2, and 3 (Brigham and Gapenski 1980,
Electrical World 1981, Resources for the Future 1980). Opportunity
costs are generally considered to be reflected in the return on equity
(Resources for the Future 1980). The objective is to identify a capital
charge rate with which to discount future costs.

81

7.2 The Discount Rate

Costs and benefits incurred in the future are usually discounted to arrive at their real values. A sum of money paid or received at a future date is not worth its current value if either or both of two conditions exist. First, if the receiver of the future money must delay present consumption or gratification while waiting for use of the money, then he must be paid some additional amount at the future date in order to remain equally well off. If not, he is better off to take the money now and either spend it or invest it and earn interest. The amount necessary to keep the individual indifferent between present versus future receipt of the money reflects his level of tolerance toward delaying gratification. It can be calculated as a percentage rate. The rate varies from individual to individual. The average of all individuals rates reflects society's time preference for consumption. If society as a whole becomes less patient, the rate will rise.

A second determinant of the discount rate is the time value of money as influenced by inflation. If inflation rises at 10 percent annually, then the sum of money to be paid or received at some future date must be increased at a 10 percent rate compounded per year in order to maintain its current purchasing power.

The overall discount rate should compensate for delayed gratification and inflation. If their respective rates are five and ten percent, then the discount rate is 15 percent. Thus the real value of the $100 today is equivalent to $115 paid or received one year from now.

When money is lent for a profit, the lender expects a total return sufficient to cover four items: the principal, delayed gratification, inflation, and profit. In the example in the preceding paragraph, if the desired rate of profit is five percent, then the total discount rate is 20 percent. The lender of $100 today will demand $120 one year from now, setting an annual interest rate of 20 percent.

The discount rate is termed such because of another way of looking at the above principles. When a nominal sum of money is to be paid or received at a future date, the question becomes: "What is the present value of the future money?" When the appropriate discount rate has been derived from the cumulative impacts of delayed gratification, inflation

and profit, the present value can be determined by answering another
question: "What amount of money put into savings today at the rate of
interest equal to the discount rate will equal the future sum of money
on the date on which it is to be paid or received?" Thus, at a discount
rate of 20 percent, the present value of $120 one year from now is $100
because $100 could be invested today at 20 percent interest and yield
$120 in one year.

Corporate firms, public agencies and financial institutions use
formulas of varying complexity in choosing discount rates most reflec-
tive of their particular situation and financial structure. Concerning
utility plants, studies by the American Gas Association (1981) and ICF,
Inc. (1980) used the following formula to estimate the appropriate real
discount rate:

$$RDR = \left[\frac{(\% \text{ DE}) (\% \text{ BY}) + (\% \text{ EQ}) \left[\frac{\frac{1}{E}}{P} \right]}{INF} \right] - 1 \qquad 7.1$$

where:

$$RDR = \text{Real discount rate}$$
$$\% \text{ DE} = \text{Percent debt capital}$$
$$\% \text{ BY} = \text{Percent utility bond yields}$$
$$\% \text{ EQ} = \text{Percent equity capital}$$
$$\frac{E}{P} = \text{Earnings to price ratio on utility stocks}$$
$$CTR = \text{Corporate tax rate}$$
$$INF = \text{Inflation rate}$$

Table 7.1 presents the values used in Equation 7.1 by the American
Gas Association; ICF, Inc. and the ICAAS   Model. The real discount
rate used in the ICAAS Model is thus 10.6 percent, rounded to
approximately 11 percent.

Table 7.1. Values of Economic Variables to Determine Real Discount Rate

Model Variable	AGA	ICF	ICAAS
% DE	50%	30%	50%
% BY	12%	12%	12%
% EQ	50%	70%	50%
$\frac{E}{P}$	6.8	10.89*	6.8
CTR	46%	46%	46%
INF	8.8%	5.5%	8.1%
RDR	9.9%	9.5%	10.6%

*Interpolated from partial data in addendum to report.

7.3 Shutdown Periods

The time necessary to make the conversions involved in various oil backout options has two potentially large impacts on the total costs of conversion. Lost revenues during shutdown periods are opportunity costs that can temporarily reduce profits. Their magnitude depends on the length of shutdown time and the availability of reserve capacity through brokerage systems or firm interchanges.

A second and equally important impact of the time factor in making oil-backout investments is inflationary rises in capital costs. If planning and carrying out the retrofitting or converting takes longer than one year, then the conversion costs in this study must be adjusted for inflation. Lengthy lead time on procuring precipitators and baghouses may extend the time frame for completion of some conversions past one year.

Both impacts of shutdown periods can be most accurately determined by individual plants. Sufficient data on these aspects were not available for analysis in this study.

7.4 Future Fuel Prices

Fuel prices directly affect the viability of conversion scenarios. ICF (1980) assumed future fuel prices would rise proportional to the overall inflation rate, i.e. no real annual increase or decrease in fuel prices. This may be a weak assumption. Natural gas and petroleum prices in the producer side of the United States economy have risen at annual rate considerably greater than the general inflation rate since 1977; i.e. 28.5 percent, 30.5 percent and 11.3 percent, respectively (ICF 1980, Ellerbrock 1981). Coal prices have risen at a yearly rate of 6.1 percent since 1977 (ICF 1980, Ellerbrock 1981). The following rates of real annual growth, as derived from the TERA-II Model (American Gas Association 1981, See Chapter 6), are used in this study: coal 4.6 percent; petroleum 4.1 percent; natural gas 6.1 percent (approximately).

7.5 Payback Period

In addition to estimating the percentage return on a capital investment and the net benefits generated by an investment, a third decision criterion is becoming increasingly popular: the payback period, i.e. the length of time it will take for the present value of accumulated net benefits to equal the total capital conversion costs. Payback period analysis is frequently used in assessing the merits of energy conservation investments.

Table 5.4 presents estimates of the payback period of conversion options on a typical 200 megawatt plant. These were calculated by dividing the total capital conversion cost by the differential operation costs per year. The unknown shutdown costs would lengthen the payback periods somewhat probably to a greater extent for conversions with the higher capital costs.

CHAPTER 8

ASSESSMENT OF THE GAS COAL-CONVERSION OPTION FOR FLORIDA

8.1. Modification of Florida's 10-Year Site Plan to Accommodate Gas-
       Coal Displacement of Oil

The possibility of converting utility oil boilers in Florida should
be considered in relation to current plans for the Florida Electric
Generation System. The major aspects of the 1980 composite Ten-Year
Plan, as compiled and prepared by the Florida Electric Power
Coordinating Group (FCG 1980) are summarized in the first and second
blocks of Table 8.1 and in Figure 8.1. The projections incorporate a
4.6 percent annual growth in summer and winter peak demand and a 4.3
percent growth of net energy for load. Twenty-two major generating
units are planned to begin commercial operation in the 1980-1989
period. Of these 2 are oil fired, 1 is nuclear, and 19 are coal fired,
4 of them out of state. The FCG document acknowledges that there are
difficulties in forecasting caused "by the threat of oil embargos, major
strikes by organized labor, the continually increasing intervention of
governmental and special interest groups into the planning, operating,
and rate-making process, and the unknown effect that financing future
facilities will have on our economic system. Also, it continues to
remain unknown at this time what the effects of the recently passed
National Energy Act will be."

Since the issuance of the 1980 FCG document in September, 1980,
many developments on the national scene have confirmed the appropriate-
ness of the caveats quoted above. The new national administration has
already implemented relaxations of the Powerplant and Industrial Fuel
Use Act (PIFUA) of 1978. For example, in view of improved gas supply
projections DOE Secretary, James Edwards, has recently ruled that in the
future natural gas will not be considered a primary fuel when it
provides less than 25 percent of the total energy of the boiler fuel

86

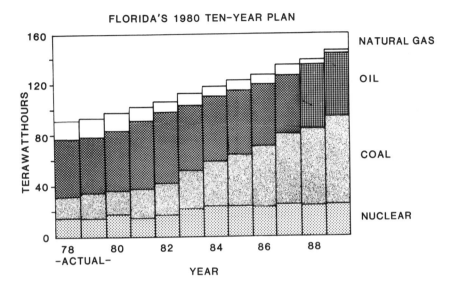

Figure 8.1 Forecast of Net System Generation by Fuel Type by
Florida Electric Power Coordinating Group

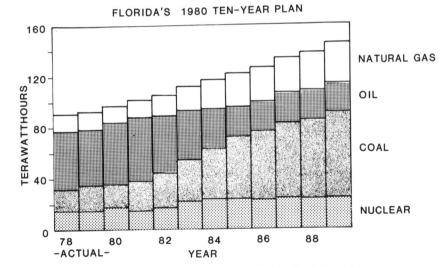

Figure 8.2 Forecast of Net System Generation by Fuel Type by
ICAAS

Table 8.1 Percent Generation and Annual Fuel Use in ICAAS's
10-Year Site Plan and a Gas-Coal Oil Displacement Modification

	1980 (FCG)		1989 FCG		1985 ICAAS		1989 ICAAS	
	%	Quant	%	Quant	%	Quant	%	Quant
Oil	49.4	78.3	35.3	83.3	18.4	38.7	18.6	45.8
Coal	19.3	9.3	46.7	30.0	41.1	21.6	41.6	25.5
Gas	13.1	132.4	1.0	16.0	21.3	268.8	21.6	318.1
Nuclear	18.2	196.0	16.8	269.0	19.2	241.9	18.2	269.0

- Converting 60% of oil boilers in State to gas/coal (70% coal, 30% gas)
- 1980 and 1989 FCG figures from 1980 Ten-Year Plan State of Florida

Note: Oil quantities are in million barrels.
       Coal quantities are in million tons.
       Gas quantities are in billion cubic feet.
       Nuclear quantities are in trillion BTU.
       All other sources are less than 0.5% and less than 0.0025
       trillion BTU.

use. This ruling would appear to accommodate our 25:75 percent gas:coal
scenario. There are also indications that the act itself will probably
be amended since gas backout provisions of PIFUA appear to be under
particularly heavy discussion. An amendment or ruling that would accommodate the peaking use of natural gas to 40 percent would probably be
favorably considered since it would largely displace distillate oil, a
higher priced and scarcer resource.

It is suggested that the State of Florida give consideration to a
modified electrical generation system expansion plan that includes
continued use of natural gas rather than strict conformance with the gas
backout provisions of PIFUA. For example, the third and fourth blocks
of Table 8.1 give the ICAAS projections for 1985 and 1989 based upon the
same energy expansion assumptions (4.3 percent) as in the FCG document
but modified to include conversion of oil boilers to gas-coal burning.

The ICAAS projections represent the amount of fuel necessary after converting suitable oil boilers in the State of Florida to coal-natural gas use. The number of megawatts convertible, 12,678, represents 60 percent of the total number of megawatts generated with oil in the State of Florida in 1980. These projections were calculated by taking the total BTUs of oil in 1980 (78.3 million barrels of oil x 6 million BTU per barrel) multiplied by 60 percent. This number of BTUs was then shifted to the gas and coal categories using a 70 percent coal, 30 percent gas ratio. For consistency with the composite FCG plan and individual power companies 10-year site plans (1981) a 4.3 percent annual energy growth rate was used to project these new figures to 1985 and 1989. The percentage generation figures are calculated according to the percent generated of BTUs per quantities of fuel projected. In contrast with the FCG plan for 1989, the ICAAS 1989 plan incorporates a substantial reduction in oil use. Figure 8.2 illustrates this plan. The economic benefits of such a reduction would be very large. It should be noted that with stronger conservation measures which would hold Florida's energy growth rate below 4.3 percent per year the economic benefits would be greater yet.

8.2. Natural Gas Transmission Capabilities

The question arises as to whether the natural gas transmission system can accommodate the increases projected from our analysis. The system as it stands now is fairly complete and there are only two major areas without service, Fort Myers and Key West. This causes a problem with the Fort Myers plant because of our classification of it as a B candidate whereas Key West is in the D category. Therefore, if Fort Myers were to convert to coal/gas it would need a gas line hook-up. The distances of the Florida power plants from the natural gas pipelines can be found in the alternate fuel column (DOE 1978) of Table 5.1. A summary of the remote power capacity (in megawatts) divided in distance groups and also categorized with our coal convertibility scale is given in Table 8.2.

Table 8.2   Power Capacities vs Distances from Pipeline
Distance to Pipeline (MCR)*

Category	On Site	<10	10-20	>20 (Miles)
A	1145	1573	-	-
B	5247	2716	1156	563
B$_1$	2645	2601	1023	506
B$_2$	2426	-	133	-
B$_3$	176	115	-	57
C	124	119	-	-
TOTAL	6516	4408	1156	620

*51% of the total MCR for categories A, B & C have pipelines on site;
35% are within 10 miles; 9% between 10 and 20 miles; 5% are greater than
20 miles from a pipeline.

The cost of a hook-up for utility plants some distance from a
pipeline is difficult to estimate. Most gas companies will provide the
hook-up if a long-term purchase guarantee is made and an entitlement is
arranged through Federal and State agencies.

The ICAAS projections (See Table 8.1) show an increase in natural
gas use from 132 billion cubic feet (BCF) in 1980 to 268 BCF in 1985 and
318 BCF in 1989. Whether or not this increased gas use can be handled
by the existing pipelines in Florida is a valid question. Through a
telephone conversation (August 18, 1981) with Mr. Harry Stout, Vice
President-Market Services for Florida Gas Transmission Company (FGT), it
was determined that the projected figures could be met by expansion of
its existing pipeline system. Presently, FGT can deliver about 735
billion cubic feet per day to the State of Florida. To double the
capacity of the existing system would require necessary support from the
purchasers in Florida, regulatory approvals, and a significant amount of
construction which could require a lengthy period, say, four years of

lead time before the capacity would be available to Florida. The capacity to meet ICAAS's natural gas projection could only be made available if the proper incentives exist to the pipeline's customers and if expeditious action were received on any application by FGT to expand the capacity of its facilities.

8.3 Analysis of the Gas-Coal Conversion Option

We have used the foregoing economic analysis techniques and quantitative economic inputs from Chapters 5, 6 and 7 to arrive at a summary of the potential economic benefits of gas-coal conversion of Florida's oil boilers. Table 8.3 is a summary of projected costs of converting steam turbined oil boilers to coal-natural gas units. The summary involved 29 power plants with a total of 58 oil boilers (i.e., candidates rated groups A, B, and C). The synopsis involves the conversion of 12,678 megawatts of electrical generation. The comparison concentrates on the period 1981 through 1995 with all costs presented in constant 1980 dollars.

Following the capital costs column and data column 8 in Table 8.3, (TOTAL, under coal-natural gas heading) the costs on which the comparison is based were calculated as discussed in the following subsections:

8.3.1 Total Conversion Costs

Total conversion costs of 1428.8 were calculated by summing conversion costs of columns A, B, and C. This is the expenditure needed for complete conversion of the 58 boilers and complete handling of coal and natural gas. Because we believe that economies of scale are present in converting these plants, a scaling factor was used in the calculations.

The calculation is the cost per kilowatt (Table 5.2) multiplied by the individual boiler maximum continuous ratings (MCR), then multiplied by a ratio of 200/MCR raised to the power of 0.2. This is equivalent to using a scaling factor of 0.8. The cost per kilowatt changes according to the classification (i.e., A, B, or C).

8.3.2. Total Annualized Capital Costs

The total conversion cost of 1428.8 was annualized for 15 years using an 11 percent discount rate (explained in Chapter 7). In other words, if 198.7 million dollars were put into a fund every year for 15 years at an 11 percent rate of return, this sum would equal the sum necessary for reconversion in 1995. Therefore, 198.7 is the cost of conversion per year.

Table 8.3 Economic Conversion Summary Table (Millions of Dollars)

Capital Costs	Oil Base Case				Coal/Nat'l Gas			
	A	B	C	TOTAL	A	B	C	TOTAL
MCR (MW)	2720	9711	247	12,678	2720	9711	247	12,678
Total Conversion Costs	--	--	--	--	240	1136	52.8	1428.8
Total Annualized Capital Costs (15 yrs. @ 11%)	--	--	--	--	33.4	158	7.3	198.7
Annualized O & M Costs (Above Oil O & M)	--	--	--	--	5.8	22.4	.9	29.1
Annual Fuel Costs Oil (@ $30/bbl)	1232	4399	112	5,743				
Coal (@ $40/ton)					288	1026	26	1,340
Gas (@ $2.60/Mcf)					192	686	18	896
	1232	4399	112	5,743	480	1712	44	2,236
Total Annualized Fuel Costs (15 yrs) Oil @ 4.1% NGR	1035	3695	94	4,824				
Coal @ 4.6% NGR					252	899	23	1,174
Gas @ 6.1% NGR					191	683	18	892
TOTAL	1035	3695	94	4,824	443	1582	41	2,066
Total Annualized Capital, O & M & Fuel Costs	1035	3695	94	4,824	482	1762	49	2,294
Production Costs per KWH	7.2¢	7.2¢	7.2¢	7.2¢	3.4¢	3.5¢	3.8¢	3.4¢

8.3.3. Annualized O & M Costs

Operation and maintenance costs (O and M) were generated from the
figures in Table 5.4. For example, 500,000 was used in generating the
cost per kilowatt of coal-natural gas units. All operation and
maintenance costs are those costs that exceed O and M for the base case
of oil boilers. Thus, 29,100,000 per year represents those O and M
costs above costs of operating and maintaining an oil boiler.

8.3.4. Annual Fuel Costs

In determining the annual fuel costs $30 per barrel, $40 per ton
and $2.60 per million cubic feet were used as the base 1980 prices. The
first two prices were used in analyses by Burns and Roe (Philipp 1980)
and to facilitate comparisons are retained here. The gas price was
identified on the basis of Florida's purchases. Inflation factors were
included using the analyses in Chapter 6. A fuel content of 6 million
BTU per barrel of oil, 24 million BTU per ton of coal and 1 million BTU
per thousand cubic feet were used. A conversion efficiency of 33% was
also used in generating total BTU necessary for fuel use in generating
12,678 megawatts of power.

The total yearly BTU converted is calculated using

$$\text{MCR} \times (1/.33) \; \frac{100,000,000 \text{ BTU}}{29.3 \text{ MWH}} \times \frac{365 \text{ days}}{\text{year}} \times \frac{24 \text{ hr}}{\text{day}} = \text{BTU/year} \quad 8.1$$

The calculation for obtaining total annual fuel cost is

$$\frac{(\text{BTU's/yr})(\% \text{ use of fuel})}{\text{Fuel BTU content}} \; (\text{price of fuel}) = \text{Annual Fuel Cost} \quad 8.2$$

A 70/30 conversion factor of coal/gas has been used.

8.3.5. Total Annualized Fuel Costs

The data generated from Chapter 6 gave us the projected growth
rates for oil, coal, and natural gas. Again, with an inflation rate of
8.1 percent the real net growth rates of oil, coal, and natural gas are
4.1 percent, 4.6 percent, and 6.1 percent respectively.

Using equation 6.14:

$$TAFC = \frac{(1 + g)}{gn} [(1 + g)^n - 1](L)(AC)$$

where:

TAFC = Total Average Fuel Cost
g = net growth rate
n = number of years
L = load factor
AC = annual fuel costs

we obtain 2066 as the annualized total cost of fuel needed to utilize 12,678 megawatts of capacity for 15 years assuming a 60 percent load.

8.3.6. Total Annualized Capital O & M and Fuel Costs

The total annualized costs are simply a summation of total annualized capital plus operation and maintenance costs plus total annualized fuel costs. 2294 million dollars is, therefore, the total expenditure necessary per year to convert to and prorate 58 coal-natural gas boilers.

8.3.7. Production Costs

The production costs were calculated on a cost per kilowatt hour basis. One multiples the total BTU per year by an efficiency rating of 0.33 to get the total BTU generated. The total BTU generated were then divided by 3412 (amount of BTU equivalent to 1 KWH) to obtain the total kilowatt hours generated. The cost per kilowatt is then simply the total annual cost divided by the total kilowatt hours generated.

These figures are deflated since the operation and maintenance costs in the base case are taken as zero. Their relative importance, though, is clear. They show in relative terms the savings in costs of production that are achieved by conversion. The economic summary clearly shows the importance of conversion, due to the high cost per BTU of oil as demonstrated in the fuel calculations.

The savings of almost 2.5 billion dollars (4.824-2.294) obviously
has beneficial ramifications for the consumer of electricity. The
savings are not directly transferable but should generate a strong
downward pressure upon the price of electricity.

8.4. Sensitivity Analysis of the Gas-Coal Conversion Option

When one considers all the possible economic and political happen-
ings that could occur that would change the price of fuel and other
factors entering our analysis, a sensitivity analysis is necessary.
Fuel price changes could be crucial, due to the overwhelming percentage
of total cost they generate in our oil to gas-coal conversion analysis
(Table 8.3). At 30 dollars a barrel we have a total annual fuel cost of
4.82 billion dollars. Assume that for political (deregulation) or
economical (cost of production and handling) reasons the cost of gas
doubled to $5.20 per thousand cubic feet. The total annual cost of fuel
as calculated would then increase from 2.07 to 2.96 billion dollars.
This is still well below the oil cost. Even if coal prices were to
double along with a doubling of gas prices the cost (4.13 billion
dollars) is still well below the cost of oil. Then if you add in the
conversion costs of 198 million dollars plus O & M costs of 29 million
dollars, there still is a savings in converting to coal-gas of 464
million dollars. Of course, if the oil price went up the savings by a
gas and coal conversion would go up correspondingly.

In view of the importance of coal prices we have examined possible
trends in coal prices in another way. Average delivered coal prices by
percent sulfur content for each month from February 1980 to February
1981 were extracted from the DOE Cost and Quality Reports (1980, 1981)
for National, South Atlantic Region, and Florida. The National average
prices for the intervals showed a maximum in the 1.0-1.5 percent sulfur
region, while Florida peaked into the 1.5 percent - 2 percent sulfur
region. South Atlantic data showed the highest price for coal with less
than 0.5 percent sulfur and decreased fairly well with higher sulfur
content which would be expected. We then fit the monthly data by our
nonlinear equation, discussed in Chapter 6 with t = 1/12 for each
month. The results showed an annual inflation rate for the year for

National = 9.9 percent, South Atlantic = 7.0 percent and Florida = 11.3 percent. "No-sulfur" coal prices were estimated to be 1.35, 1.72, and 2.09 dollars per million BTU respectively. The Florida monthly analysis agreed quite well with the yearly data shown in Figure 6.1. The additional premium for lower percent sulfur content coal was less on the National scale (β = .014) than for South Atlantic (β = .097) or Florida (β = .093) during the designated period. The range of prices suggests that coal energy costs will remain well below oil energy costs.

8.5. Simplified Economic Comparisons of Other Oil to Coal Conversion Options to the Gas Coal Conversion Option

We have not carried out a complete analysis for the other forms of oil to coal conversions. However, by extrapolating from Table 5.2 and 5.4, we estimate that the saving with the coal-oil mixture approach would be about 0.2 to 0.3 billion dollars. This is down an order of magnitude from the gas-coal mixture approach. The problem with coal-oil mixture is that it has only been possible to work to a 50-50 mixture by weight of coal and oil. Because of the lower calorific energy per lb. of coal, this corresponds to a 30-40 percent use of coal energy and a 70-60 percent dependence upon oil energy. While some conversion costs are saved, the continued heavy reliance on oil, the highest priced fuel greatly reduces the benefits of this conversion. The gas-coal conversion gets its much greater savings by virtue of its large (70 percent of energy) reliance on the cheaper fuel (coal).

Again projecting from Chapter 5, we estimate that the dual coal-oil approach without flue gas desulfurization should yield an annual saving of the order of 1.8 billion dollars. The gain with respect to coal-oil mixture is due to the much greater reliance upon coal energy. The dependence upon oil during peak seasons, when coal use is impeded by the derating problem is quite significant, however. In addition to paying a higher price for oil energy, it is necessary to pay an energy or cost penalty to continuously heat the residual oil which would otherwise congeal to the consistency of asphalt.

Conversion to dual coal/oil capability with flue gas desulfuri-zation (FGD) should yield a saving of about 1.2 billion dollars. On the

other hand building a new coal boiler with FGD should yield a savings of
1.0 billion dollars, which is not much less. It should be noted that in
these calculations we have assigned a price of $120 per kilowatt for FGD
(See Table 5.2) However, because of space and load limitation $200-$300
per kilowatt has been suggested as a realistic cost of a retrofit
installation of FGD (AGA 1981). Along with the added capital conversion
costs there is also an energy loss associated with FGD with the need to
reheat the flue gases following scrubbing. Thus a flue gas
desulfurization requirement would be a major deterrent to oil-to-coal
conversions, and if FGD can be avoided without compromising air quality,
it would occasion considerable savings.

Conversion to coal and water is the next most serious alternative
for oil backout now under consideration because it leads to a pumpable
fuel. On the other hand the optimum coal/water mix is now estimated at
50/40. This suffers from energy costs of boiling a lot of water and the
energy costs of pulverizing the coal to small particle sizes ($\sim 10\mu$).
We roughly estimate that the annual Florida savings would be about 1.5
billion dollars when consideration is given to the cost of peaking units
to replace the power for the lost capacity.

Conversion to gas-coal and water would maintain the pumpability of
the water/coal mixture but would use natural gas to share the energy
load, and assist with the water boiling. The higher flame temperature
might facilitate the use of normal pulverized coal sizes. We estimate
that the overall savings in this case would be about 2.0 billion dollars
per year.

Finally it might be noted that capital costs of conversion represent a relatively minor annualized cost factor (see Table 8.3). Thus
even if capital cost estimates come out higher it would not greatly
lower the gain associated with gas-coal conversion. In view of the
great potential saving of oil imports perhaps a national program of low
interest loans for oil to gas-coal conversions might be arranged.

CHAPTER 9

AIR QUALITY IMPACTS

9.1 Quantitative Measures of the Problem

The major potential environmental impact of oil-to-coal conversions
can arise because of the vastly greater ash content of all coals and
from the somewhat greater sulfur content of many coals. Table 1.5 gives
the characteristics of several of the major fuel groups, along with
their energy release, SO_2 emission and ash emission in
lbs. Figure 1.6 is a nomogram which may be used to carry out the
calculation of emission levels for any fuel when the percentage by
weight and the calorific values (in 10^3BTU/lb) are specified. The lines
illustrate the major fuel groups listed in Table 1.5. Note that the
high sulfur oil example leads to greater emissions than two of the coal
groups. From the SO_2 emissions standpoint the conversion to coal can
have a positive or negative environmental impact.

One of the primary purposes of the simultaneous use of gas with
coal in retrofitting oil boilers is to meet state implementation
standards (SIP) for sulfur dioxide without the use of capital, expense,
and space intensive flue gas desulfurization (FGD). The strict emission
standards in New England proved to be a critical matter in the American
Gas Association (1981) study of select gas use. The situation is more
complicated in our study of oil to gas-coal conversions in Florida
because the SIPs in various regions of the state vary widely. This is
illustrated in Table 9.1, which gives the emission limits for various
regions in Florida according to current regulations of the Florida
Department of Environmental Regulation. Four extra columns have been
added to the table representing the percent of natural gas (by energy)
which must be burned with various grades of coal to meet the regional
SIPs. Using an academic approach the coals are here graded from an

98

Table 9.1 Florida Emission Limiting Standards and Percentage Natural Gas Needed to Meet Emission Limits for Various Grades of Coal. Emissions are given in the Unit σ = 1 lb of SO_2 per million BTU Heat Input

Stationary Sources	Emission	Percentage NG A 1.2σ	B 2.4σ	C 4.2σ	D 6.0σ
		Emission potential ÷			
FOSSIL FUEL STEAM GENERATORS					
A. New sources burning					
1. Liquid fuel	0.8 σ maximum two hour average	33.3	66.7	80.9	86.7
2. Solid fuel	1.2 σ maximum two hour average	0	50.0	71.4	80
3. Gaseous fuel	--				
B. Existing Sources					
1. Liquid fuel					
a. Duval County North of Heckscher Dr. except Jacksonville Electric Authority Northside Stations	2.5 σ maximum two hour average	0	0	40.5	58.3
b. Jacksonville Electric Authority's Northside Stations	1.98 σ	0	17.5	52.9	67
c. Jacksonville Electric Authority's Southside and Kennedy Stations	1.10 σ	8.3	54.2	73.8	81.7
d. All other sources in Duval County	1.65 σ	0	31.2	60.7	72.5
2. Solid fuel					
a. Hillsborough County, Tampa Electric Co. Francis J. Gannon Station Units 5 and 6 and Units 1-4 upon conversion to solid fuel	Units 1-6 total not more than 10.6 tons per hour of sulfur dioxide on a weekly average and a maximum unit length of 2.4 σ on a weekly average.	0	0	42.9	60
b. Hillsborough County, Tampa Electric Company's Big Bend Station, Units 1, 2 and 3	Units 1, 2 and 3 total not more than 31.5 tons per hour of sulfur dioxide in a three hour average but not to exceed a two hour average emission of 6.5 σ	0	0	0	0
c. Escambia County Gulf Power Co. Crist Units 4, 5, 6 and 7	5.90 σ	0	0	0	1.7
d. All other areas of the state	6.17 σ	0	0	0	0
e. Hillsborough Co. including Tampa Electric Co. Gannon Units 1-4 prior to conversion to solid fuel, and Hooker's Point Station	1.1 σ	8.3	54.2	73.8	81.7
f. Escambia County, Gulf Power Co. Crist Units 1, 2 and 3	1.98 σ	0	17.5	52.9	67
g. Escambia County Monsanto Textiles Units 1-8	57.5 tons in any 24 hour period				
h. Manatee County, Florida Power and Light Company's Manatee Station	1.1 σ	83.	54.2	73.8	81.7
i. Leon & Wakulla County Tallahassee's A.B. Hopkins and Purdom Stations	1.87 σ	0	22.1	55.5	68.8
j. Dade, Broward and Palm Beach Counties, Florida Power and Light Company's Cutler Units No. 4, 5 and 6, Ft. Lauderdale Units No. 4 and 5, and Riviera Units No. 1 and 2.	1.1 σ, except in the event of a fuel or energy crisis	0	0	34.5	54.2
k. All other areas of the State	2.75 σ	0	0	34.5	54.2

environmental viewpoint according to the scheme $A(1.26\sigma)$, $B(2.46\sigma)$ $C(4.26\sigma)$ and $D(6.06\sigma)$ (σ denotes lbs/10^6 BTU).

The last four columns of Table 9.1 suggest that reasonable proportions of natural gas would make it possible to meet state SIPs if grade B coal were used. This observation is more clearly demonstrated in the last four columns in Table 4.2 which gives the percentage of natural gas needed when using the various academic grades of coal for various uniformly spaced hypothetical SIPs. It should be clear that the simultaneous use of natural gas serves best primarily when high environmental standards are reflected in the chosen SIPs. This presents somewhat of a dilemma because apart from compliance coal (the A group), the other major coal groups in the United States are not readily brought to 1.2σ compliance by using natural gas mixing alone.

Since the Florida Sulfur Oxide Study (FSOS) (Wilson et al. 1978), many of Florida's oil boilers have been licensed to release up to 2.75σ. However, very few of these plants are utilizing these limits, not only for reasons of environmental protection, but also to facilitate complying with particulate standards. In view of the fact that Florida is already having acid rain problems, and is probably having attendant visibility degradation, the opportunity exists upon conversion to coal to make a significant reduction in SO_2 emissions, or at least to hold the line at the current emission or ambient levels.

The regulatory procedures, which will govern oil to coal conversion, are still a matter of uncertainty. If the conversion is carried out voluntarily it presumably must conform to Best Available Control Technology (BACT). For solid fuels, this corresponds to 0.1 lb particulates per million BTU input for a maximum 2-hour average. Despite the large ash content of most coals, the efficiency of precipitators or bag houses is so high that particulate emission compliance is not a serious problem. The SO_2 limit of 1.2σ for a maximum 2 hour average, however, is a matter of considerable controversy particularly because to meet BACT usually entails the installation of flue gas desulfurization (FGD). In retrofitting, FGD would add very considerable capital costs to an oil to coal conversion. One obvious regulatory solution would be to waive BACT but to require emissions not to exceed preconversion oil

emissions. Here a problem arises as to whether the limits on oil are
the actual historic limits or the regulatory limits. The actual are
generally lower than the regulatory limits allowed in Florida on a
region to region basis.

Following the Florida Sulfur Oxide Study (Wilson et al. 1978),
regional emissions standards were established, which were largely based
upon emissions which could be accommodated within national ambient air
quality standards (NAAQS) and Florida's ambient air quality standards
(FAAQS). The NAAQS are based upon human health effects. The national
secondary standards, which are Florida's primary standards, also attempt
to protect plants, animals and materials. It might be mentioned that
both NAAQS and FAAQS do not give consideration to acid rain and
visibility, which are beginning to become problems for Florida. Let us
next consider what techniques are available for maintaining FAAQS and
emission standards corresponding to BACT or at least a reasonable
available control technology (RACT)

9.2 Methods of Reducing Sulfur Emissions

(1) High Quality Coal

The simplest control technology is to use compliant coal i.e. coal
which upon combustion produces less than 1.2σ. This is available from
Central Appalachian and Western United States (ICF, Inc., 1980). These
coals are not in short supply and the price penalty at this time is not
very high (about 12 percent increase per one percent of SO_2 decrease).
It might be expected, however, that compliant coal will increase in
price at a faster rate than noncompliant coal.

(2) Coal Cleaning

A second technique to reduce sulfur emissions is to use physical
cleaning (PC) either on site or at the mine. This is a well-developed
technology and one which usually has a considerable benefit in relation
to costs (E.P.A. 600/2 1976, 600/9 1977). Among the benefits of PC are:
(a) the increased heat content of the cleaner coal, (b) savings in
transportation costs, (c) savings in costs of ash disposal, (d)
pulverizing cost savings, (e) boiler and related equipment savings.
Table 9.2 lists the properties of a sample of coals before and after

cleaning. The contrast in sulfur is largely determined by the percent of pyritic sulfur in the coal, which is largely removeable by PC. It should be clear that considerable reduction in ash content and SO_2 production can be achieved by coal cleaning at reasonable cost ($\sim$\$0.1 per 10^6 BTU).

Table 9.2 PROPERTIES OF SAMPLE COALS BEFORE AND AFTER CLEANING

COAL LOCATION	% SULFUR		RATIO	% ASH		RATIO
	RAW	CLEAN	C/R	RAW	CLEAN	C/R
Sullivan County, Indiana	1.87	1.11	.59	10.5	7.8	.70
Cambria County, Penna	2.40	1.01	.42	11.4	6.7	.59
Harrison County, Ohio	2.30	1.26	.55	10.4	4.8	.46
Clearfield County, Penna	0.85	0.45	.53	9.3	7.0	.75
Preston County, W Va.	2.24	1.25	.56	18.5	11.9	.64
Armstrong County, Penna	2.53	1.09	.43	13.0	7.2	.55
Jefferson County, Ohio	2.82	2.03	.72	9.8	6.0	.61
Vigo County, Indiana	1.54	0.40	.58	12.0	7.7	.64
Garrett County, Maryland	2.37	1.60	.68	13.8	8.8	.64
Franklin County, Illinois	1.12	0.95	.85	14.8	7.1	.48
Greene County, Penna	3.45	2.20	.64	11.4	8.1	.71
Marion County, W. Va.	3.80	2.16	.57	11.0	5.9	.54

The study that provided the data in Table 9.2 concluded that the costs to meet emission standards by physical cleaning followed by flue gas desulfurization was substantially lower (up to 60 percent) than the use of FGD alone. It might be surmised that to comply with BACT, or a form of RACT without FGD, the benefits of PC would be even greater.

Tables 9.3 and 9.4, contain information (DOE/EIS-0038, 1979) on coal production and coal resources for Demand Region IV (Alabama, Florida, Georgia, Kentucky, Mississippi, North Carolina, South Carolina, and Tennessee). All coal mined in Region IV (Table 9.3) comes from the southern end of the Appalachian coal basin except for Western Kentucky coal and Illinois basin deposits. Most of the coal is high-volatile bituminous. Notable exceptions are the lignites of southern Alabama and Mississippi. Region IV demonstrated coal reserves are shown in Table

Table 9.3: Demand Region IV Coal Production and Quality, 1975 (millions of short tons)

State	Total Production	% Ash Average	% Ash Range	% Total Sulfur Average	% Total Sulfur Range	% Pyritic Sulfur	Btu/lb Average	Btu/lb Range
Alabama	22.644	9.5	1.2-21.7	1.3	0.4-4.9	0.70	13,000	11,310-15,140
Georgia	0.060	5.8	2.4-9.0	0.8	0.7-1.0	NAb	14,475	14,398-14,628
Kentucky	143.613	12.8	3.5-39.0	2.8	0.6-5.1	1.24	12,602	8,543-14,200
Tennessee	8.206	7.9	1.9-23.8	1.2	0.4-5.8	0.40	13,670	10,940-15,000
Regional Total	174.523	12.1	1.2-39.0	2.5	0.4-5.8	1.13	12,802	8,543-14,200

aNo coal was mined in Florida, Mississippi, North Carolina, or South Carolina in 1975.
bNA--Not available

Table 9.4 Demand Region IV Demonstrated Coal Reserve Base as of January 1, 1974 a,b (millions of short tons)

State c	Reserves by % Sulfur Content <1.0	1.1-3.0	>3.0	Unknown	Total Reserves
Alabama	624.7	1,099.9	16.4	1,239.4	2,981.8
Georgia	0.3	0	0	0.2	0.5
East Kentucky	6,558.4	3,321.8	229.5	2,729.3	12,916.7
West Kentucky	0.2	564.4	9,243.9	2,815.9	12,623.9
North Carolina	0	0		31.7	31.7
Tennessee	204.8	533.2	156.6	88.0	986.7
Regional Total	7,388.4	5,519.3	9,716.4	6,904.5	29,541.3

aBituminous coal and lignite.

bDemonstrated reserves are those coal deposits which have been measured or geologically projected to a high confidence level and are a sufficient quality and volume to be economically mined at the time of the determination.

cNo measurable reserves in Florida or South Carolina. Mississippi has some lignite reserves.

SOURCE: U.S. Department of Energy, Fuel Use Act, DOE/EIS-0038, April 1979.

9.4 with coal deposits less than 1829 m (6000 ft) deep and hypothetical resources estimated at 162,357 million short tons. These tables indicate that clean and cleanable coals should be available to Florida for reasonable prices for many years. The same statement should be applicable to most of the other states.

(3) Gas-Coal Mixtures

Gas-Coal burning provides another simple procedure for reducing the SO_2 emission per 10^6 BTU. (Schlesinger 1980, American Gas Association 1981) The reductions are easy to calculate, since the relationship is linear. For example a 25-75 (by energy) use of gas and coal reduces emissions per 10^6 BTU to 0.75 of the coal emissions. If gas is in very good supply one might go to 40-60 (or multiplication by 0.6).

(4) Burner or Boiler Desulfurization

The fourth technique is to use burner or boiler scrubbing. This has been developed to a high degree in the so called Fluidized Bed Combustion method. For standard boiler or burner arrangements the reduction achieved to date by this technique is not comparable to flue gas desulfurization but the costs are much lower. The Trimex process, (Milner 1980) which uses bentonite or calcium montmorillonite on the boiler surface, which achieves a reduction factor of say 0.6 is an example of such a boiler process. The injection of powdered limestone into a burner could also lead to a reduction of this magnitude at low costs (Rawls, 1981).

(5) Flue Gas Desulfurization (FGD)

The most direct and efficient process for removing SO_2 is to chemically treat the flue gases to remove SO_2. Efficiencies for removal go to 90 percent, and the standards of BACT have largely been determined by what is achievable with FGD. Unfortunately to install a FGD unit requires space, which may be difficult to find when retrofitting an existing oil boiler. This, plus the capital and operational costs of FGD, has greatly dampened the interest in this technology in oil-coal conversions.

The possibility of achieving BACT without requiring FGD exists if one uses a combination of the measures (1), (2), (3) and (4) described above.

Let us consider an example of the effects of the three simple reduction measures upon a middle (C) grade coal e.g. 2.5 percent sulfur at 12,000 BTU. Unabated this coal would emit 4.2σ. With R(PC) = 0.6 this reduces to 2.5σ. With gas-coal burning in the proportion 25 to 75 which would also facilitate the conversion with a minimum derating R(NG) = 0.75 we reduce the emissions to 1.9σ. With simple burner scrubbing R(BS) $\approx$ 0.6 the emission goes to 1.1σ. Thus with an average (C) grade coal three relatively inexpensive measures achieve BACT.

9.3 A Reasonable Available Control Technology

In approaching the new question of oil boiler to coal conversion and the very new oil boiler to gas-coal conversion question it is desir- able, in view of the economic urgency of the matter, to establish a reasonable available control technology (RACT). This RACT should at- tempt to minimize annualized costs of conversion, operation and main- tenance, and fuel, and at the same time minimize the SO_2 emissions and its attendant external costs. The latter should include (a) health effects of atmospheric products of SO_2, (b) plant and wildlife effects (c) effects on lakes of acid rain--particularly the tourism and commer- cial fishing industries and (d) visibility degradation as it affects both residents and tourists.

The four measures described in the previous section make it possi- ble to reduce SO_2 emissions to arbitrarily low levels even below the 1.2σ now agreed upon as representing BACT. Each measure, however, adds a cost factor to the eventual price of the electricity to the consumer. Hence a knowledgeable citizen may not wish to have SO_2 reduction mea- sures carried out beyond what is reasonable. The question remains as to how to quantify reasonable.

Benefit cost analysis is one approach to quantification. Several recent attempts have been made to carry out detailed benefit cost analy- ses in environmental energy questions of this nature (Lave and Seskin 1973, ICAAS 1978, Loehman et al. 1979). While responses by specialists to such work have been favorable it appears unlikely that there will be general public acceptance of any new attempts at benefit costs analyses in the short time frame available for addressing the oil-coal conversion

question. Perhaps, however, a knowledgeable public would accept a simpler approach of minimizing a "convertability index" such as a simple
geometric mean index

$$I = \left[\frac{C}{C_{std}} \cdot \frac{Q}{Q_{std}}\right]^{1/2} \qquad\qquad 9.1$$

where C is the cost of energy delivered to the public and C_{std} is the
cost as calculated for an agreed upon conversion scenario. The standard
conversion scenario, could be taken as conversion to coal with FGD, with
derating, but with retention of oil use for peak periods. Here Q is the
actual SO_2 emissions and Q_{STD} is 1.2σ (BACT). The public or represen-
tatives are essentially given the choice of tolerating some $Q > Q_{std}$ if
the cost of the electricity following conversion is substantially lower
than that for a standard conversion scenario.

Thus if a utility proposes a specific oil-to-coal conversion it
must also show that the index I for the proposed conversion is smaller
than or in the neighborhood of unity. For example if $Q = 2.4\sigma$ is
proposed (i.e., $Q/Q_{std} = 2$) then the proposed conversion must indicate a
substantial reduction in consumer electricity costs say $C \approx 1/2\ C_{std}$.
For this example the composite index would still be 1.00 which would be
regarded as qualifying for RACT.

Other algebraic formulas for I could be used to weight the cost-
emission trade-off differently: The geometric mean appears to be the
most reasonable of several tested by the author. The algebraic indices
would be relatively easy to apply, since estimates of C and Q would
naturally be included in any conversion application. The conversion
costing methods described in this study and in references could be
applied to explore alternatives to help identify the most reasonable
conversions.

It would be difficult at this time to spell out specific rules for
the most RACT for oil to coal conversions. Nevertheless it does appear
that the simultaneous use of four simple measures: (1) clean coal, (2)
physical cleaning, (3) gas-coal mixing and, (4) boiler or burner scrubbing together with any sensible oil to coal conversion should yield net
indices competitive with FGD.

CHAPTER 10

NATIONAL ASSESSMENT FOR OIL BACKOUT

10.1 Overview

In Chapter 8 a boiler by boiler study of Florida's utility oil boilers indicated that we could displace 80-90% of the oil currently used by Florida Utilities which for 1980 (See Figure 1.2) corresponds to about 0.2 million barrels of oil per day. To gain some perspective as to the national potential for the displacement of oil by natural gas and coal it is helpful to have an overall national energy picture both by energy supply and by energy use. Table 1.7 gives such a picture as expressed in quads per year which corresponds to 0.46μ (μ denotes a million barrels of oil per day oil equivalent). This simple conversion is helpful in gaining an overview of the potential for gas-coal displacement of oil in the United States.

Table 1.7 indicates that our overall energy supply by oil in 1980 corresponds to 32.7 quads per year or approximately 15μ. Of this 14.2 quads (6.6μ) represents imported oil and 18.2 quads (8.3μ) represents domestic oil. It should also be noted that the transportation use of oil corresponds to 18.6 quads (8.5μ). Since a liquid fuel at normal temperatures provides the most compact and expedient way of storing chemical energy, liquid products distilled from oil are the most natural fuels for transportation. Accordingly, unless a major break through in achieving more efficient vehicle engines occurs there is little likelihood for major oil saving in transportation. Thus, a reasonable immediate goal would be to direct our domestic oil production towards fulfilling domestic transportation needs. The recent development of ways to upgrade residual oil into higher transportation hydrocarbons will facilitate this goal and at the same time will apply further economic pressure to finding alternatives to residual oil use in utility and industrial boilers.

107

Figure 1.1 shows that on a national scale electrical generation consumed 1.4μ of oil. This is a reasonable target for gas-coal displacement, and even if only 70 percent effective would constitute a great economic saving to this country (1.0μ). A more detailed examination of the utility oil stream will be made in Section 10.2.

The industrial use of oil, 3.4μ, constitutes an even more tempting alternative fuels target. In Section 10.3 we will examine the potential of reducing this stream. In Section 10.4 we will look into the possibility of reducing the 1.4μ of commercial oil use and the 1.6μ of residential oil use.

10.2 Utility Oil Displacement

A summary of Utility Boilers in the United States designed for Oil Firing Currently Burning Oil ($\geqslant$ 50 megawatts) has been compiled by Jamgochian, et al. (1980, see their Table 2). The data for the total 63,279 megawatt capacity representing 245 units and 107 stations are subdivided by Standard Federal Region and state. California with 21,601 megawatts, Florida with 12,656, and New York with 6,560 have about 2/3 of the large utility oil boiler capacity in the United States. The next largest users with over a thousand megawatt capacity are Illinois (2,849), Pennsylvania (2,482), Massachusetts (2,060), Virginia (1,727), Arizona (1,616), Michigan (1,395), Connecticut (1,248), Texas (1,246), Maryland (1,236), and New Jersey (1,165). The remaining 37 states have a total of 5,438 megawatts.

To go into details as to the retrofitability of these oil boilers would require an extensive study which not only gives consideration to the detailed characteristics of these boilers but also the ambient air quality standards of the varying regions and the state and local emission standards. These emission standards vary widely, somewhat like those for the different locations in Florida described in Chapter 9. In many instances because of much higher population concentrations and industrialization, the emission standards are more strict than BACT (e.g., Boston Area SIP = 0.286). This implies that to retrofit an oil boiler in such a region without flue gas desulfurization would require

the maximum use of the four measures described in Section 9.2. Thus it
would be essential to start with a high quality coal, physically clean
it, use natural gas with it, and use burner or boiler scrubbing. The
cost of such measures would vary on a region-by-region basis. However,
until detailed analyses are made it is not unreasonable, to estimate the
annualized savings (over a 15-year period) for the U.S. as a whole on
the basis of the detailed Florida study. The ratio of U.S./Florida
large oil boiler capacities is

$$R_{mw} = 63,279/12,678 = 5.0$$

is the convenient projection factor. On this basis, the 2.5 billion
dollars annualized savings projected for Florida proportions to a U.S.
savings of 12.5 billion dollars. The savings in oil, according to R_{mw},
can also be calculated by this proportion. Our Florida study indicated
that the conversion of 12,678 megawatts to gas   coal is accompanied by
a saving of about 0.20 million barrels of oil per day or 73 million
barrels of oil per year. Ratioing to the U.S. as a whole, one projects
a saving of 1.0 million barrels of oil a day or 2.2 quads per year.
This level of imported oil could thus be displaced by the combined use
of domestic gas and coal. While the projected U.S. annual savings to
the consumer is 12.5 billion dollars, the savings to the U.S. as a
country might be assigned double this value. Thus from the viewpoint of
the nation as a whole at this time, the important thing is that the
dollars involved stay at home and do not become a part of the outflow of
dollars associated with our negative balance of payment problems.

10.3 Industrial Oil Back-out

While the savings to the electricity consumer and to the country in
the displacement of utility oil are quite dramatic, these savings are
small compared to the potential displacment of oil use by industry.
This is illustrated in Figure 1.1 which shows that in 1980 the industrial use of oil was 3.4 million barrels per day 2.4 times the utility
use. To determine accurately the amount of this oil which could be
displaced by gas and coal would really require a new detailed analysis

since our Florida study did not examine industrial boilers. However many of the considerations which apply to small utility boilers are directly applicable to large industrial boilers. Furthermore, there have been a number of studies carried out on a national basis addressing the question of "Replacing Oil and Natural Gas by Coal in Industry" (Harris et al., 1976; Roach et al., 1978; Mayster, 1979; Dyck et al., 1979; Cohen, 1980 a, b; Shaw, 1979; Ehrlich et al., 1980). Much of the substance of these studies can be utilized to help provide approximate answers to the question of "Replacing Oil by Coal and Natural Gas," the topic of this monograph.

Existing oil/gas fired large industrial boilers (>100 million BTU/hr) consume about 1.7 quads and small boilers (<100 million BTU/hr) consume about 1.2 quads. Existing process heaters are capable of 4.2 quads and feedstocks based upon liquefied gas, oil, and natural gas consume 5.1 quads (Cohen 1980a).

On the basis of our Florida Utilities study, it is not unreasonable to assume that 70 percent of the 1.7 quads $\approx(0.8\mu)$ consumable by large industrial boilers can be displaced by gas and coal. This displacement corresponds to 1.2 quad (0.55μ).

The small boiler question is more difficult. Table 10.1 (Cohen, 1980a, Table 3.9) provides a summary of Industrial/Commercial Boilers prepared by PEDCo Environmental Inc. The table indicates that of the total of 1.85 million million BTU/hr capacity, coal is the primary fuel in only 13 percent, whereas 29.5 percent is residual oil, and 10.2 percent is distillate oil. Natural gas provides 47.2 percent of the primary fuel which when broken down by size of boiler corresponds (in million BTU/hr) to 10 percent (50-100), 12 percent (25-50), 10 percent (10-25), and 15 percent (<10). The large proportion of natural gas use in industrial boilers might greatly facilitate the gas-coal conversion option since this is an existing gas stream that can probably be diverted somewhat to the advantage of the boiler owners, in saving of fuel costs, and the country as a whole, by providing a source of gas to be used with the coal to displace residual and distillate fuel oil.

Leaving the small gas boilers (1 million BTU/hr) unchanged it would appear that by replacing larger gas boilers with 50-50 gas-coal mixtures

Table 10.1 Capacity Distribution of Small Industrial Boilers by Size and Fuel Type (10^6 BTU/hr)

Primary Fuel Type	<0.4	0.4-1.5	1.5-10	10-25	25-50	50-100	Total	%
Coal	4,100	14,260	25,250	26,280	75,980	95,200	241,070	13.0
Residual	10,300	42,520	117,840	98,660	154,120	121,650	545,090	29.5
Distillate	6,400	22,740	63,180	47,170	35,010	14,660	189,160	10.2
Natural Gas	26,400	88,830	199,740	164,700	210,810	182,800	873,360	47.2
TOTAL	47,200	168,350	406,010	336,810	475,920	414,390	1,848,680	99.9
Percent	2.6	9.1	22.0	18.2	25.7	22.4	100.0	

SOURCE: PEDCo Environmental, Inc., "The Population and Characteristics of Industrial/Commercial Boilers, prepared for Environmental Protection Agency, May 1979, Tables 2-9, 2-11, 2-13.

Table 10.2 Coal, Gas, and Fuel Oil Consumption in Process Heat Applications in 1974 by Industry (10^{12} BTU)

Industry	Coal	Natural Gas	Distillate Fuel Oil	Residual Fuel Oil	Total Fuel Oil
Food		94			
Textiles		26			
Paper		80		60	60
Chemicals		436	31	42	73
Petroleum refining		799	29	196	225
Cement, lime, glass, etc.	258	688	55	48	103
Primary metals	6	687	41	281	322
TOTALS	264	2,790	156	627	783

SOURCE: Energy Consumption Data Base, prepared for FEA by EEA, Inc., Arlington, Virginia, June 9, 1977, from DOE/EIA-10547-01 "Technical Feasibility of Coal Use in Industrial Process Heat Applications," draft report prepared for DOE by EEA, Inc., Arlington, Virginia, May 22, 1978.

we could reduce the total proportion of natural gas use to 32 percent. A 40-60 gas-coal mix should probably suffice for the distillate oil boilers and a 30-70 gas-coal mix for the residual oil boilers. Calculation of such a reshuffling of fuel use indicates that in the industrial boiler sector (>10 million BTU) coal can be shifted from 13 percent to take over about 50 percent of the load and natural gas can be increased slightly from 47 percent to 50 percent. The displacement of oil in the industrial oil boiler sector would then be complete.

The prospects in the process heat industrial sector are approximately as good as in the boiler industrial sector. Table 10.2 (Cohen, 1980 Table 3.11) shows the proportion. In the cement, brick, and lime industry coal not only has the advantage of lower cost but the high radiation characteristics of a coal flame can be of technological advantage. Shifts to coal in other process uses would be very dependent upon the possibility of cleaning up the particulate and gaseous emissions of coal. To the extent that gas-coal burning can facilitate this clean up so would it be possible to make further inroads in the displacement of oil in process heating applications.

The feedstock use of oil for synthetic fiber and for plastics manufacturing represents another possible area of savings. Here coal could serve as alternative feedstock thus displacing oil and releasing natural gas for higher use. In addition biomass and botanochemical crops which were the original feedstocks before oil became cheap are also available (Zaborsky 1981, Buchanan and Duke 1981). By focusing biomass studies towards this goal the U.S. might make more rapid strides towards further oil importation reduction. (Mace and Ellis, Smith and Dowd 1981)

In summary it would appear relatively easy to displace 2μ of the 3.4 μ currently used in the industrial sector and by working hard at it we could possibly achieve a displacement of 2.6μ.

10.4 Commercial and Residential Area

The 1980 use of oil in the commercial area was 1.4μ and in the residential area 1.6μ. The fractions of these fuels which can be displaced by coal alone has been estimated rather pessimistically because of the greater technological and environmental problems which coal presents. However, little attention has been given to the possible displacement of oil by gas and coal.

One aspect of these sectors is that natural gas is extensively used, equivalent to 1.1μ, in the commercial sector and 2.4μ in the residential sector. Thus by relatively little reshuffling it should be possible to facilitate the displacment of oil energy by coal and gas energy in these sectors if technological developments can cope with reliability, supply, and environmental problems. Since the sizes of boilers or furnaces in the commercial sector fits in a continuous fashion with the smaller end of the industrial sector, one might extrapolate the displacement potential here in an approximate way. The development of fluidized bed combustion systems has opened new doors towards small scale coal systems. We believe gas use with fluidized beds would enhance this trend. This belief is reinforced by the recent development of a tri-fuel stoker fuel burner-boiler system which achieves high efficiency and clean burning in boilers under 30,000 pounds of steam per hour or 30 million BTU/hr. (Combustion Service and Equipment Co., 1981). An example of this system in the 10 million BTU per hour range led to average stack emission in the 0.25 $lb/10^6$ BTU range. Systems are currently available as low as 1.4 million BTU/hr. A stoker system is used for the coal whereas blowers are used to overfire the coal bed with an oil or gas flame. This system, in itself, could have a large impact on the small industrial market, and the commercial market. There is also a possibility that pulverized coal-natural gas system can be developed for the small industrial market although this is contrary to conventional wisdom.

The possibility of going smaller yet into the residential market itself has barely been explored although recently a stoker fired central heating system using anthracite coal has been marketed by Tarm, Denmarks largest boiler manufacturer (Energy Daily, December, 1, 1981). We believe the gas-coal displacement option also opens up a great variety of possibilities. To a large extent these possibilities might draw upon the extensive privately and publically supported work that has been done on coal gasifiers. Table 3.4 gives a listing of the coal gasifiers in various stages of development (Miller and Lee, Chapter 6, CBI, 1980).

One sees that the low BTU gasifiers are quite far along. A number of producer gasifiers work best locally and emit relatively de-ashed flue gases suitable for combustion in a second stage chamber. The difficulty in these cases is that the flame temperature obtainable with 150 BTU/ft^3 feet is several hundred degrees Fahrenheit below that achievable with oil or natural gas. On the other hand, if this BTU gas is mixed with natural gas in reasonable proportions. one obtains a mixed gas with a rather high BTU fuel as good or better than that achievable with the more sophisticated and costly gasification processes. By using a well insulated delivery system the high temperature of the combustible gases and air should be sufficient to produce a flame temperature equal to that of oil in oil boilers or furnaces. Accordingly, it is believed that the small size barrier to coal use which exists in combustion folklore can be overcome by using gas and coal simultaneously together with the lastest coal gasification technology.

As to how much oil can be displaced by gas and coal in the commercial field, it is believed that 60 percent should be possible for a potential savings of 0.8μ. In the residential field these new gas-coal systems would be applicable to apartment houses, condominiums, and clusters of residences. Perhaps 35 percent is an achievable goal, corresponding to a savings of 0.6μ. Little extra gas would be needed in commercial and residential transitions since the large quantities of gas use in these sectors could be redistributed to good advantage. Furthermore, conservation moves are very effective in this sector and solar energy is very suitable for water heating needs. Thus several measures are available to displace natural gas for a higher use as a facilitator of coal combustion.

Table 1.8 gives conservative (low), neutral (best), and optimistic (high) estimates of oil displacement levels which might be achieved in the utility, industrial, commercial and residential sectors by vigorous programs directed towards gas-coal replacement of oil. Note that if the optimistic estimates were realized we would practically have achieved energy independence.

CHAPTER 11

RESEARCH AND DEVELOPMENT NEEDS

Burning gas-coal mixtures is not a research topic that a scientist
would naturally choose in the pursuit of truth and knowledge. It is too
complicated a combustion system to be of great interest from a fundamental standpoint. However, it is a reasonable topic, somewhat outside the
pale of topics currently under study in major combustion programs, which
might appropriately be pursued in connection with an attempt to solve an
important national problem. In this case, the problem is to reduce our
large negative balance of payments associated with the importation of
oil from the OPEC Cartel.

11.1 Existing Research Programs

Resource geologists have for some time predicted the exhaustion of
the oil and natural gas reserves of the world and particularly the
United States. The Arab oil embargo of 1973 and the subsequent price
increases by the OPEC Cartel from around 2 dollars per barrel in 1972 to
around 32 dollars per barrel in 1981 have focused considerable attention
on our depletion problem. The economic problems associated with these
price increases have prompted major expansions in our research and
development programs directed toward finding alternative energy
sources. The earlier national efforts tended to concentrate on non-
fossil sources such as breeder reactors, controlled fusion, solar
energy, geothermal energy sources, ocean thermal sources, etc. As our
immediate economic problems became more severe, it has become generally
recognized that coal represents the only large domestic energy resource
which this country can draw upon quickly to help correct its economic
problems (Energy Research Advisory Board, 1980; Landsberg, et al. 1979;
Office of Technology Assessment, 1979; National Academy of Sciences,
1980; ICAAS, Coal Burning Issues, 1980).

115

Much of the emphasis of the coal related research and development work has been directed towards finding ways of mitigating the environmental impacts usually attributed to the use of coal. In addition to oil displacement the multibillion dollar synthetic liquid and gaseous fuels program initiated by President Carter was, in part, motivated by this emphasis. The development of clean coal solids, using physical and chemical coal cleaning, and flue gas desulfurization systems are essentially attempts to clean up the coal act.

11.2 Current Oil Backout Programs

A federal program has been underway for several years "encouraging" the restoration to coal use of boilers originally designed for coal burning but converted to oil during the the days of cheap oil (Fuel Use Act, 1978). With the near completion of this program attention naturally falls upon the question of retrofitting to coal, boilers originally designed for oil. Conventional wisdom in the United States had it that this is not economically feasible. For example the Regional Utility System of Gainesville, Florida, which placed an order for a 240 megawatt oil boiler in 1974, suffered severe cost penalties when at the beginning of the construction phase, it completely changed over to a 240 megawatt coal system under federal mandate. ICAAS's detailed involvement in the environmental impacts of utility boilers along with its consideration of coal use goes back to this period (ICAAS, 1975).

The commissioner and staff of the Kwinana Utility Station near Perth, Australia faced with exponentially increasing oil prices and apparently unconstrained by U.S. conventional wisdom, (Kirkwood et al., 1978; Treadgold, Engineers Australia 1979; Energy Commission News, March 1979) chose to retrofit two 200 megawatt oil boilers still in the construction phase to dual coal/oil use. Since then planning to retrofit oil boilers to burn coal directly has also become an acceptable engineering activity in the United States.

11.3 Pumpable Coal Based Substitutes for Oil

While the dual coal/oil solution is also under consideration in Tucson, Arizona (CE 1980), the trend at this time in this country is toward the development of pumpable coal based substitutes for oil. The coal/oil mixture approach which has this feature has thus far received the greatest attention (Cook 1980). However, it would appear now that the economic benefits of this approach do not significantly exceed costs and this approach will probably begin to lose the priority that it has been given in U.S. R & D efforts.

The next highest priority for investigation by the U.S. Department of Energy and also by the Electric Power Research Institute appears to be the coal-water approach. This is another pumpable fuel arrangement which has the advantage that the pulverizing of the coal and the fuel mixing can be carried out outside the immediate space of the oil boiler itself--thus simplifying retrofitting problems. More importantly, it reduces immediate capital costs which is a major problem of the utility industry at this time (Brigham and Gapenski, CBI 1980).

Certain questions, however, can be raised concerning the coal-water approach which remain to be answered. In the first place, the fact that a good quality coal by itself is a marginal oil boiler fuel, except possibly for a generously sized oil boiler designed with coal retrofitability in mind, would be exacerbated by the large water content of the injected coal-water slurry. Presently a slurry consisting of 40 percent water and 60 percent coal is viewed as optimum. Such a mixture would have a decreased calorific content, of the order of 8,000 BTU/lb. which is comparable to lignite. The vaporization and heat capacity loss (between inlet water temperature and flue gas temperature) corresponds to about 480 BTU/lb which corresponds about to a 5 percent energy loss in hot water vapor out the stack. More importantly, a large derating might be expected (see Fig. 4.1) since such a mixture could only be burned at a low power density (usually called space heating rate). Some compensation to avoid this lower power density could be accomplished by using a higher degree of pulverization, i.e., 90 percent through 325 mesh but this is costly in pulverizer energy and pulverizer machinery wear. Other compensations possibly can be made by incorporating new

ways of increasing the power density such as used in gas turbine reactors. Another problem might be the adverse effect upon flue gas
inerting systems (Smoot et al. 1980) designed to minimize pulverizer
fires and explosions. Water vapor is also known to promote the
conversion of sulfur to sulfuric acid and the extra corrosion potential
of a coal-water slurry could be deleterious to flue gas inerting system
and other systems of the retrofitted oil boiler.

The use of a pumpable slurry of 60 percent pulverized coal, 30 percent methanol and 10 percent water has been discussed (Capehart et al.
CBI 1980a). The calorific context of methanol is 9800 BTU/lb which
makes it a much more favorable liquid coal carrier than water, from a
combustion standpoint. Whether its cost, in $/BTU makes it competitive
and whether the derating problem is substantially mitigated remains to
be determined.

11.4 Gas-Coal Burning Configurations

The economic and/or environmental advantages of the gas-coal approach with respect to the coal-water approach must ultimately be determined by demonstration projects which, of course, will require time and
money. To determine how best to spend this time and money it would be
helpful to sketch the general characteristics of possible gas-coal
burning arrangements. In retrofitting, design options are limited by
the space availability, foundation capabilities, detailed boiler design,
etc. Many aspects of these problems have already been overcome in the
Kwinana project and in the proposed Irvington conversion of the Tucson
Gas, Electric Light and Power Co. where "shoe horn" approaches were
used.

11.4.1 Direct Gas Coal Firing

Figure 11.1 illustrates the main features of coal boiler arrangement apart from the fuel input and burner. Obviously a retrofit should
move as far as possible toward a coal designed boiler. Figure 11.2
illustrates, on an enlarged scale, the gas-coal feed system and a burner
which if space is available could be placed immediately at the boiler
face. If space is not available, the coal pulverizing components shown

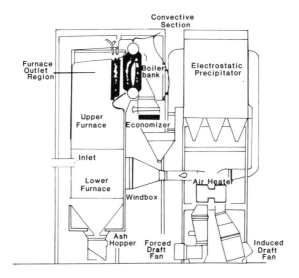

Fig. 11.1 Schematic Diagram of a Pulverized Coal-Fired Electric Utility Boiler (Adapted from Hardesty and Pohl 1979, Sandia Laboratories Report 78-8804 and Coal Study Report, Cooper et al. 1981).

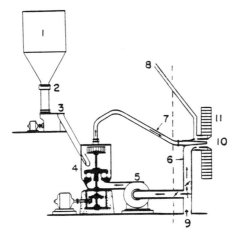

Fig. 11.2 System for Direct Firing of Pulverized Coal with Gas (1) Coarse Coal Bin (2) Gate (3) Feeder (4) Pulverizer (5) Fan (6) Secondary Air (7) Pulverized Coal and Primary Air (8) Gas (9) Heated Air (10) Burner (11) Boiler. (Adapted from Figure 3.14(a) Essenhigh, 1979 Chapter 3, Coal Combustion in Coal Conversion Technology, Wen and Lee eds. Addison-Wesley, 1979).

in Figure 11.2 could be located at the nearest available space-e.g. some 100-200 feet distant. The combination would be similar to the dual coal/oil configuration except that the retention of the oil capability would be unnecessary since during peak power needs the proportion of natural gas would simply be increased, say to 40 percent. Thus the gas-coal approach avoids the necessity of maintaining the residual oil system in a state of readiness (by keeping the oil warm) or reliance on the use of expensive peaking units. The disadvantage of this direct gas-coal injection system is the large amount of ash injected into the oil boiler. The required boiler modifications to accommodate this ash should be similar to those used at Kwinana and planned for at Irvington.

11.4.2 Two Stage Gas-Coal Firing

A second gas-coal option is illustrated in Fig. 11.3 which shows input equipment which replaces that shown in Fig. 11.2. This equipment includes a first stage combustion system designed to remove as much of the coal ash as possible and to inject into the boiler for after-burning a fuel rich hot mixture of CO, coal volatiles, and unburned natural gas. The boiler itself, fed by secondary and tertiary air, would burn this mixture along with additional natural gas. If boiler front space is limited, the equipment illustrated in Fig. 11.3 could be placed immediately outside the boiler area. Blowers and high temperature insulated pipes would be used to transport the gaseous products for afterburning in the oil boiler with natural gas enhancement. The dis-tance between the pulverizer and first stage combustion chamber-ash separator might be used to advantage to allow further means for separat-ing any residual ash before injecting the hot gaseous combustion prod-ucts into the boiler. By working below ash melting temperatures it should be possible to minimize slagging problems in both combustion chambers and in the transport tubes and also NO_x problems associated with atmospheric nitrogen. By working the first chamber at fuel rich conditions NO_x problems due to chemically bound coal nitrogen can be substantially reduced (Hall and Cichanowicz 1980, CE 1981 p. 13-9).

Another advantage of the first stage combustion chamber approach is that it affords the possiblity of carrying out some SO_x scrubbing such

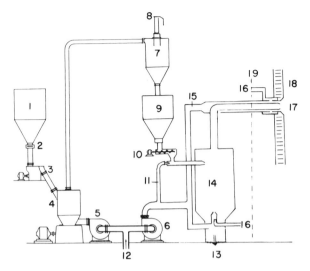

Fig. 11.3 Systems for Two Stage Pulverized Coal-Gas Firing (1) Coarse Coal Bin (2) Gate (3) Feeder (4) Pulverizer (5) Pulverizer Fan (6) Burner Fan (7) Cyclone Separator (8) Air Vent (9) Pulverized Coal Storage Bin (10) Pulverized Coal Feeder (11) Primary Air (12) Air Supply (13) Ash Outlet (14) First State Combustion Chamber-Ash Separator (15) Secondary Air (16) Gas Inlets (17) Coal-Gas Burner (18) Boiler (19) Pipe Extension if needed. (Adapted from Fig. 3.14(b) Essenhigh, 1979, Chapter 3, Coal Combustion in Coal Conversion Technology, Wen and Lee eds).

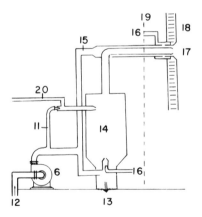

Fig. 11.4 Schematic System for Two Stage Coal-Water-Gas Firing. Here (20) Represents Input from Coal-Water Pipeline and all Remaining Numbers are the same as in Fig. 11.3.

as referred to in Section 9.2. If such suppression techniques could also serve to convert the ash into a useable product such as a fertilizer, this would enhance the benefits of the conversion in relation to the costs.

Finally, it should be mentioned that the first stage gas-coal combustion chamber approach affords an opportunity to start with a coarser pulverized coal since by the use of vortex motion the residence time in the burning zone could be prolonged to equal the burnout time of the coarser particles.

11.4.3 Gas-Coal-Water System

In arrangements based upon the gas-coal-water approach, the pulverizer and coal-water mixing system could be as far away as the nearest port facility. In direct firing of the coal water slurry a dual fuel burner similar to that used in gas-oil burners could be used in an arrangement similar to that illustrated in Figure 11.2. The natural gas facilitates the water evaporation, increases the calorific value of the fuel thereby increasing the achievable power density beyond the capabilities of the coal-water fuel.

A first stage combustion chamber-gasifier-ash separator such as is illustrated in Figure 11.4 might be used in the gas-coal-water approach. The first stage would effectively be a generator of H_2 + CO, i.e. producer gas. The overall arrangement would minimize the boiler derating problem and provide other advantages such as a capability for directly using the coarse (80 percent through 14 mesh; granulated sugar sized) particles used in coal slurry pipelines. Such an approach would greatly reduce the capital costs required for an individual oil boiler retrofit in exchange for increased fuel costs. The approach could be implemented in a metropolitan area even before long distance coal pipelines are built.

The two gas-coal configurations and the gas-coal-water configuration just discussed suggest a specific applied research and development efforts which will be discussed next.

11.5 Research Needs

Section 11.4 visualizes three broad types of systems for retrofitting oil boilers to burn coal with natural gas or to burn a coal-water
mixture with natural gas. If we accept the development of such systems
as a short term goal where should we apply research efforts to hasten
the implementation of such conversions? Four research topics stand out.

1. Studies of the ash formation process in interacting gas-coal
   flames.
2. Studies of the characteristic of afterburning flames which can
   be created from various gas-coal primary flames.
3. Studies of the influence of additives in the primary stage upon
   the production of SO_x, NO_x and upon other characteristics of the
   afterburning flame.
4. Studies of the influence of the water in the gas-coal-water
   approach upon the ash formation process, the characteristics of
   the afterburning flame for various inputs and for various types
   of additives.

11.5.1 Ash Formation in Interacting Gas-Coal Flames

To some extent there already is a background of scientific research
on gas-coal flames, in studies of the burning of coal having high volatile content. Several recent reviews of coal pyrolysis have appeared
[Horton (1979), Malte and Rees (1979), Wen and Dutta (1979), Essenhigh
1979] as a general background which can serve for gas-coal studies. The
primary types of experiments needed for applications are experiments
which resemble those arising in practical gas-coal combustion systems.
The experimenter would control the quality of the pulverized coal, its
fineness and flow field, and the natural gas and air inputs. Heating
rates and final temperature, chemical and spectral emissions of the
resultant flame should be measured. Flame temperature might be controlled with an insulating jacket or a water jacket with various flow
rates.

Spectroscopic and photometric measurements and ash collection could
be used to establish the influence of the higher heating rates expected

in gas-coal flames upon devolatilization, and the combustion of the
char. Existing kinetic models of pyrolysis should be generalized to
allow for the externally introduced methane. The pyrolysis equations
should be coupled with transport equations to arrive at a model of the
temperature, volatiles and combustion product evolution under realistic
coal, gas and air input rates.

Fly ash formation in pulverized coal combustion is a very complex
phenomena (Ulrich et al., 1976; Flagan, 1977). The initial heating and
devolatilization of coal particles and char burnout usually results in a
skeletal char structure with mineral inclusions. At about 50% burnout
the hot molten ash inclusions tend to fuse into an agglomeration which
accelerates in the last 25% of char burnout. Later in the burnout the
char structure fractures, resulting in fly-ash fragments. How will this
fly-ash evolution change when the coal particles are burning with natural gas? The higher temperature achievable should, through Arrhenius
factors, increase all reaction rates and speed up the entire chain of
combustion processes. On the other hand competition for oxygen might
have a slowing effect. The influence of the proportions of coal, gas
and air would be important to understand. The influence of flow field
arrangements such as vortex motion which extends residence time should
be studied and quantitative descriptions should be developed. Ash
collection and classification with various flow field arrangement should
be carried out in conjunction with modeling to infer the parameters of
the fly-ash formation processes.

11.5.2 Afterburning Studies

A major simplification in the retrofitting of oil boilers could be
accomplished if a first-stage combustion chamber can be developed to
serve a multiple role: (1) pyrolyze the coal; (2) promote the ash
formation process and separate out a large fraction of the ash; (3)
deliver a fuel rich mixture of intermediate gaseous combustion products
to a second stage combustion chamber for afterburning; (4) provide for
the introduction of SO_x and NO_x suppressing additives.

The spectral output of afterburning flames which is expected to be
richer in transient species than the flame of the first stage combustion

chamber should be analyzed. Furthermore, the relationships between the particulate sizes emanating from the first stage chamber and the conditions within this chamber should be investigated by examining the spectra and the gaseous and particle emissions of the afterburning flame.

The overall objective of afterburning flame studies is to be able to predict the characteristics of the combustible mixture emitted by the first stage combustion chamber. A theoretical model of these relationships should be developed which would be useful in predicting the characteristics of boiler flames in practical two-stage combustion arrangements.

11.5.3 The Influence of Additives Upon Flame Characteristics and Afterburning Characteristics

Desulfurization in fluidized bed coal combustion systems has been developed to a high level of sophistication. The possiblity of carrying out desulfurization in a pulverized coal boiler or a first stage combustion chamber has received much less attention. What is needed is to investigate the influence of various flame additives upon SO_x and NO_x emissions. Of particular interest would be the use of inexpensive additives such as limestone, dolomite, or waste phosphate slime, which would transform the ash into useful commercial fertilizers or other commercial products.

11.5.4 Coal-Water-Gas Gasifiers

A primary stage coal-water-gas combustor has the interesting extra dimension of high water content in the input fuel. Under appropriate circumstances this should foster the formation of synfuel or producer gas (H_2 + CO). The intensive work which has already been carried out on gasifiers (see Table 3.4) should find applications here. Studies which bridge the gap between gasification programs without methane augmentation and the mixed gas-coal-water gasification would be very useful. Of particular interest would be whether, with the gas-coal-water combustor, one could utilize directly the coarser grind now under consideration in coal water slurries. If it can be burned efficiently this would make it possible to greatly reduce the local capital costs of conversion. Essentially the conversion would consist of the installation of several package coal-water-gas gasifiers with high temperature piping arranged to feed the output into the boiler. Ash removal facilities will still be needed but this is only about 10 percent of the problem posed by the coal facilities which are no longer needed.

11.5.5 General Studies of Gas-Solids Combustion and Ash Separation

The underlying thesis of this monograph is that burning coal with gas upgrades the coal as a combustible material to the point of being usable as a higher quality fuel such as oil. Since the United States has excellent reserves of coal but is economically suffering from insufficent oil reserves, the use of moderate quantities of natural gas in defense of the quality of American life can be viewed as constituting a higher use, perhaps even higher than the use of gas for home water heaters. In view of the more favorable gas supply picture the possibility of upgrading other solid fuels such a brown coal, lignite, peat, wood chips, bark chips, peletized urban wastes (Hanson 1980) so that each of them can replace a higher quality fuel also warrants consideration. General experimental investigations of gas-solid fuel combustion could serve a great variety of purposes. The development of a theoretical model for gas-solid combustion would provide a valuable basis for many applications not only for retrofitting existing boilers and furnaces but also for the design of new systems. First stage combustion chambers can be designed for many solid fuels.

The concept of a first stage natural gas-solid fuel combustor-ash-separator is a fruitful area for applied research. The concept has several features of a cyclone burner (BW 1978 Chap. 11) but it would appear best to design to achieve a dry ash, most of which is separated out by the techniques used in cyclone dust separators, before the hot combustion gases are injected into the boiler. On the other hand a high temperature first stage combustion chamber, like a cyclone but arranged to separate out the molten ash before injection into the boiler, might prove as efficient or more efficient for many fuels. Cool baffles in the gaseous transport system might stop small molten ash droplets entrained with the combustion gases. The air mixing techniques used in recent distributed mixing burners (Hall and Cichanowicz 1980, Rees et al. 1980, CE 1981, p. 13-9) might overcome the NO_x problem of ordinary cyclones. The new feature of mixing natural gas into the first stage combustion zones and into the boiler combustion zones provides additional flexibility which should challenge the imagination of combustion scientists and engineers concerned either with the design of new boilers or retrofitting an existing boiler.

The characteristics of the ash and its possible commercial uses will be important in the studies of gas-solids combustion and ash separation.

11.5.6 Basic Research

The foregoing specific topics are motivated by a need to reduce uncertainties in gas-coal or gas-coal-water systems envisaged in Section 11.4. To the extent that these types of systems have not been considered in the formulation of existing Department of Energy, Environmental Protection Agency, Electric Power Research Institute, Gas Research Institute or National Coal Research Institute research programs, they represent nonconventional combustion research programs. Very recently a report has been submitted to the American Physical Society by a study group on Research Planning for Coal Utilization and Synthetic Fuel Production (Cooper et al. 1981). This study constitutes a broad nonconventional view of coal utilization needs. The report is directed towards a two-fold audience: physicists who may become active in coal research and DOE sponsors and others in the "user community". The report emphasizes areas of coal research where the contributions of technical and conceptual approaches of physics would be especially relevant. The report discusses coal-oil mixtures and coal-water systems but does not consider gas-coal systems. Nevertheless, the execution of these research recommendations would greatly advance understanding of topics which would be helpful toward the development of practical gas-coal systems. Table 11.1 is a list (unranked) of the principal research recommendations of this committee.

Table 11.1   Principal Research Recommendations (unranked)
              (The American Physical Society Coal Study Group, Cooper et al. 1981)

● support for physics research in the area of coal utilization make provision for a non-negligible percentage of highly novel, even speculative, approaches to outstanding applied problems.

● presently available and developing techniques be applied to the determination of chemical structural features of coals, including hydroaromatic and aromatic cluster configurations, functional groups, and molecular weight distributions.

● systematic analyses of the physical properties of coal to elucidate the pore structure of coal particles including pore shapes and pore size distributions.

● strenuous efforts to develop better dynamic models of advanced coal processes.

● initiation of a carefully planned program to evaluate presently available instruments, and where necessary to develop new instruments, usable both for systematic collection of data on advanced coal processes and for incorporation into control systems for these processes.

● intensive effort on theoretical work needed to formulate problems concerned with multiphase flow behavior, and to develop innovative computer simulations of such behavior.

● an extensive research effort to understand the effects of coal conversion and combustion environments on materials of construction (alloys, refractory ceramics, composites, etc.), including long duration exposure to the process environment.

● pursuing experimental and theoretical investigations to delineate the phenomena in size reduction processes, such as grinding and chemical comminution.

● further work to bring sophisticated spectroscopic methods of solid state and surface physics to bear on systems with compositions, and conditions of pressure and temperature, as closely akin as possible to those used in practical heterogeneous metal catalysis for fuel synthesis, i.e., starting from syngas (H_2 and CO) mixtures obtained from coal gasification.

● increased emphasis on understanding and exploiting the properties of size and shape selective catalysts.

● the establishment of a sample bank of well selected, characterized, and preserved coal samples, available to research and analytical groups. This should be preceded by research sufficient to verify the chosen storage technique(s).

● establishment of a program to gain fundamental understanding of processes for coal utilization and synthetic fuel production. This would focus on the development and testing of theoretical models and of process control techniques, together with the development of instrumentation for acquisition of the data necessary for this purpose.

11.6 Development Needs

As we have indicated, applied and basic research are greatly needed to develop an understanding of the basic laws governing practical gas-coal systems. However, much is already known about the burning of high volatile pulverized coal which can be used as a basis for the design and construction of prototype gas-coal systems while basic and applied research programs are being carried out.

Thus, the development of a practical first stage gas-coal combustion chamber with efficient ash and sulfur removal features can be carried out on an empirical basis using knowledge acquired in the gasification program, by experience with cyclone burners (BW 1978) and distributed mixing burners (CE 1981). The primary objective would be to develop a gasifier-ash separator to feed clean gaseous fuel into a pre-existing boiler. The acid test of success is whether the first stage combustion products could be injected into a practical second stage furnace, process heater, kiln, or boiler. Firing a small first stage combustor (say 5 million BTU/hr) into a ceramic kiln would be particularly useful since glazes are very sensitive to residual ash. If, for example, the system could be fired with high quality stoneware without affecting the glazes, this would be indicative of success with boilers.

Experience on a moderate scale using the output of a primary combustor should provide information upon which to base retrofitting modifications for industrial level boilers or furnaces for institutional heating systems. There is a background on the use of gas and coal in limestone and brick furnaces which could also be helpful in this regard. Experience with industrial level gas-coal systems would also provide a basis for the development of scaling laws which would be useful for the design of a large range of gas-coal burner sizes. Conventional wisdom has it that pulverized coal systems are only economically competitive at large power levels (i.e., > 50,000,000 BTU/hr). On the other hand, conventional wisdom has not considered the gas-coal concept and it would be worthwhile to reexamine the cross-over point at which a pulverized coal system becomes economicaly competitive, when it is augmented by gas.

Many other technical developments would be helpful in speeding the realization of the displacement of oil by burning coal with gas. For example, if for reasons of space limitations the first stage combustion chamber is placed off-site then there is a need for high temperature insulated piping to deliver the hot fuel to the boiler. The large spatial separation between the first and second stages opens up the possibility of incorporating additional ash reduction devices enroute. The introduction of cyclone separators, loops or baffles could further reduce ash in the first stage flue gas.

There are a number of utility boilers in this country designed for coal firing but also having oil and/or gas burning capability (Kidder, Peabody and Company 1979, Plank 1981). This dual or tri-fuel capability was incorporate for start up purposes or to take advantage of the lowest price fuel and to have emissions flexibility from an environmental standpoint. These systems might serve as the immediately available test facilities for large scale testing of the gas-coal-oil displacement approach. If by burning coal with gas the steam output could be increased substantially above the rated output for coal alone (assuming no boiler tube or other limitations are exceeded) it would suggest that burning coal with gas in an oil boiler could be carried out above the coal power limit.

Many other practical developments such as control systems, safety systems, ash disposal systems, etc. will be needed to implement a gas-coal conversion program. The dual fuel character affords an extra degree of freedom which can be used very gainfully in such systems. By taking advantage of the greater ease of gas regulation and the latest developments in sensing devices and computer control systems, a world of new possibilities for automation exists. The dual fuel in itself affords a degree of redundancy which promotes reliability, which is a major consideration in the utility, industrial, commercial, and residential uses of combustion.

11.7 Interdisciplinary Research Needs

This chapter has concentrated on technical aspects of gas-coal burning since the highest priority question is whether we can "fool" a preexisting oil boiler into burning coal with gas at or near its design power capacity. Once this is established and technical objections are answered there obviously will be needs for research on the environmental-health-economic-financial-political-socio-legal and other interdisciplinary aspects of a transition from oil to gas-coal burning. In this connection the chapters on these topics in Coal Burning Issues (Urone, Fahien, Taylor et al., Bolch, Woltz and Street, Schlenker and Jaeger, Green and Rio, Brigham and Gapenski, Rosenbaum, Little and Capehart CBI 1980) also are relevant to burning coal with gas. For example Chapter 18, Federal Regulatory and Legal Aspects, is of special concern since the Public Utilities and Industrial Fuel Use Act of 1978 is an obstacle to burning coal with gas. How do we go about revising a federal law which now appears to have been based upon erroneous estimates of future natural gas production? The political side of this question and other questions such as addressed in Chapter 17, Coal and the States: A Public Choice Perspective are also very relevant.

In this latter connection the recent work by the political scientists Wildavsky and Tenenbaum (1981) entitled the Politics of Mistrust raises some very disturbing questions. They hypothesize that, for most of this century, American energy policy has been a polarized political issue in which policy differences are rooted in deep conflicts over values and hence energy policy debates cannot be resolved by better information and analysis. While this is very disheartening to the editor of this work he finds hope in their note that during the two World Wars there was enough consensus about the purposes of energy policy to mute devisiveness and mistrust. Perhaps the great economic problems which this country now faces will be viewed as the "equivalent of war". Hopefully then an objective evaluation of options based upon new information and analyses might be achieved and a consensus as to the common good might again be reached.

132 / REFERENCES

References and Author Index [Bracketed Numbers]

American Gas Association, Gas Supply Committee, <u>The Gas Energy Supply Outlook: 1980-2000</u>, Arlington, Va., October 1980 (as revised December 1981. [6, 7, 28, 31, 35, 38]

American Gas Association, TERA-II: Demand Marketplace Model DM8132, American Gas Association, Arlington, Va., July 22, 1981. [15, 73, 75, 85]

American Gas Association, Economic Comparison of Oil Scrubbed Coal, and Select Gas Use with Coal in New England Power Plants, in Energy Analysis, American Gas Association, Arlington, Va., 1981. [72, 93, 97, 98, 104]

Babcock & Wilcox, <u>Steam: Its Generation and Use</u>, 39th Edition, Babcock & Wilcox Co., New York, N.Y., 1978. [8, 44, 45, 53, 68, 127]

Bach, W., Pankrath, J. and Williams, J., <u>Interactions of Energy and Climate</u>, D. Ridel, 1980. [54]

Baumeister, T., Avallone, E. A., Baumeister, T. II, Marks's Standard Handbook for Mechanical Engineers, Eighth Edition, McGraw Hill, Inc., New York, N.Y., 1978. [65]

Bolch, W. E., Solid Waste and Trace Elements Impact, Chapter 12, pp. 231-248, in <u>Coal Burning Issues</u>, ICAAS, 1980. [131]

Bogot, A. and Sherrill, R. C., Principal Aspects of Converting Steam Generators Back to Coal Firing, Combustion, p. 10, March, 1976. [48]

Breen, B. P. and Sotter, J. G., Prog. Energy Combustion, Sec. 4, pp. 201-220, Pergamon Press Ltd., 1978. [48]

Brigham, E. F. and Gapenski, L. C., Financing Capacity Growth and Coal Conversions in the Electric Utility Industry, Chapter 16, In <u>Coal Burning Issues</u>, ICAAS, pp. 331-342, 1980. [81, 117, 131]

Brust, R. L., <u>Pit and Quarry</u>, February, 1978. [47]

Buchanan, R. A. and Duke, J. A., Botanochemicals Crops, pg. 157-179, Handbook of Biosolar Resources Vol. II, O. R. Zaborsky, Ed., C.R.C. Press, Boca Raton, Florida, 1981. [18, 112]

Cantrell, W. N., <u>Gannon Station Units 1-4 Conversion</u>, Tampa Electric Company, September, 1980. [65]

Capehart, B. L., O'Connor, J. D. and Denty, W. M., Coal Transportation, Chapter 4, pp. 71-94. In <u>Coal Burning Issues</u>, ICAAS, 1980. [118]

Coffin, K. P. and Brokaw, R. S., A General System for Calculating the Burning Rates of Particles and Drops and Comparison of Calculated Rates for Carbon, Boron, Magnesium and Iso-octane, N.A.C.A. Tech. Note 3929, 1957. [46]

Cohen, B. N., Technical and Economic Feasibility of Alternative Fuel Use in Process Heaters and Small Boilers, U.S. Department of Energy, Energy Information Administration, Contract No. DE-AC01-79EL10547, DOE/EIA-10547-01, February, 1980. [110,111,112]

Cohen, B. N., The Substitution of Coal for Oil and National Gas in the Industrial Sector, U.S. Department of Energy, DOE/EIA/TR-0253, Energy Information Administration, Assistant Administer for Applied Analysis, Washington, D.C., November, 1980. [110]

Combustion Engineering, Inc., <u>Combustion, Fossil Power Systems</u>, J. G. Singer (Ed.), Windsor, CT., 1981. [8, 45, 53, 68, 127]

Combustion Engineering Inc., Tucson Electric Power Co., Irvington Station, Attachment 1, November 17, 1980. [117]

Combustion Service & Equipment Company, CNB Tri-Fuel Boiler Brochure, 2016 Babcock Boulevard, Pittsburgh, Pennsylvania 15209. [113]

Cook, M. C., Operating Florida Power and Light Company's Sanford Plant on Coal-Oil Mixture, presented at the Seventh Annual International Conference on Coal Gasification, Liquefaction, and Conversion to Electricity, University of Pittsburgh, August 5-7, 1980. [53, 117]

Cooper, R. B. et al., Report to American Physical Society by Study Groups on Research Planning for Coal Utilization and Synthetic Fuel Production, Reviews of Modern Physics, 53 (4) Part II, October, 1981. [20, 50, 119, 127, 128]

Department of Energy, Energy Information Administration, Principal Electric Facilities, A Map of Southeastern U.S., 1978. [55, 89]

Department of Energy, Fuel Use Act: Final Environmental Impact Statement, DOE/EIS-0038, April, 1979. [102, 115]

Department of Energy, Inventory of Power Plants in the United States, Electric Power Statistics Division, December, 1979. [55]

Department of Energy, Economic Regulatory Administration, Office of Fuels Conversion, Profiles for Title I Existing Electric Powerplants, April, 1980. [72]

Department of Energy, 1980 Annual Report to Congress, DOE/EIA-0173 (80), Vol. 1, 2, 3, March, 1981. [3, 17, 31]

Department of Energy Reports, Cost and Quality of Fuels for Electric Utility Plants, DOE/EIA-0075, 1979, 1980 and 1981. [73, 76, 95]

Dyck, D. K., Demakos, P. G., White, D. C. and Cox, L. C., Comparison for Fuel and Technology Alternatives for Industrial Steam Generation Systems, MIT Energy Laboratory Working Paper MIT-EL 79-060WP, December, 1979. [110]

Ehrlich, S., Drenker, S., and Manfred, R., Coal Use in Boilers Designed for Oil and Gas Firing, presented at American Power Conference, Chicago, Illinois, April 21-23, 1980. [7, 53, 66, 110]

Electrical World, A Short Course in Utility Finance, 195, (5), 45, May, 1981. [81]

Ellerbrock, M. J., Energy Investment Savings Can Appreciate, Florida Banker, 8, (7), 25-28, July, 1981. [85]

Ellis, T. H. and Mace, A. C., Jr., Forest Research in Florida, Journal of Forestry, pp. 502-515, August, 1981. [112]

Energy Daily, December 1, 1981. [113]

Energy Research Advisory Board, 1980. [115]

Environmental Protection Agency, Sulfur Reduction Potential of U.S. Coals: A Revised Report of Investigations, EPA-600/2-76-091, April, 1976. [101]

Environmental Protection Agency, Coal Cleaning with Scrubbing for Sulfur Control: An Engineering Economical Survey, EPA-600/9-77-017, August, 1977. [15]

Eoff, K. M., Conversion of Biomass by Physical and Chemical Processes, Proceedings of Conference on Alternative Energy Sources for Florida, p. 57-62, Gainesville, Florida, December 5-6, 1979. [38]

Essenhigh, R. H., Coal Combustion, pp. 171-312, In Coal Conversion Technology, C.Y. Wen and E. A. Lee (Ed.), Addison-Wesley, Reading Massachusetss, 1979. [119, 121, 123]

Fahien, R. W., Air Pollutant Dispersion Modeling, Chapter 10, pp. 187-202. In Coal Burning Issues, ICAAS, 1980. [131]

Flagan, R. C., Ash Particle Formation in Pulverized Coal Combustion, Paper No. 77-4, Spring Meeting of the Western States Section, The Combustion Institute, Seattle, Wash., 1977. [124]

Florida Electric Power Coordinating Group, State of Florida 1980 Ten-Year Plan, September 1, 1980. [55, 86]

Florida, Governor's Energy Office, Forecasts of Energy Consumption in Florida, 1980-2000, September, 1981. [4]

Glassman, I., Combustion, Academic Press, New York, N.Y., 1977. [45]

Gold, T., and Soter, S., Terrestrial Sources of Carbon and Earthquake Outgassing, Journal of Petroleum, Geology 1 (3):3-19, 1979. [37]

Gold, T., and Soter, S., The Deep-Earth Gas Hypothesis, Scientific American, 242 (6), pp. 130-137, June, 1980. [37]

Glenn, R., Coal-Water Slurry Systems for Oil Design Power Plants, EPRI FP-1164, Combustion Processes, Inc., 1979. [10, 53]

Green, A. E. S., Scions are Fermions, A Law of Socio-Physics?, Physics Today, pp. 32-38, June, 1965. [73]

Green, A. E. S., The Middle Ultraviolet: Its Science and Technology, Wiley, New York, 1966. [5]

Green, A. E. S., A Note on Survival Curve with Shoulders, Radiation Research, 60, pp. 536-540, 1974. [73]

Green, A. E. S., Rich Phenomena in the Middle Ultraviolet, Optics News, 7, 3, p. 77, Invited Paper, National Meeting of the Optical Society of America, October 26-30, 1981. [5]

Green, A. E. S., The Fundamental Nuclear Interaction, Science, 169, pp. 933-941, September 4, 1970. [43]

Green, A. E. S., and Green, B. A. S., in Proposal by ICAAS on Mitigation of the Impact of Increased Coal Use in the Southeast (MIICUSE), to U. S. Department of Energy, p. 11, February, 1980. [5]

Green, A. E. S. and Green, B. A. S., Description of Initial Firings of Propane and Pulverized Coal at the Dragon Fire and Clay Company Facilities in Proposal to U. S. Department of Energy, dated March, 1981 (see also Green and Jones, 1981). [5, 47]

Green, A. E. S., and Jones, J. R., Gas-Coal Burning Interactions, an Addendum to IICUF, issued by ICAAS, March 15, 1981. [1, 5, 20]

Green, A. E. S., MacGregor, M. H. and Wilson, R., First International Conference on the Nucleon-Nucleon Interaction, Gainesville, Florida, Rev. Mod. Phys. 39, pp. 495-715, 1967. [43]

Green, A. E. S., and Rio, D. E., Quantitative Public Policy Assessments, Chapter 15, pp. 303-330, In Coal Burning Issues, ICAAS, 1980. [131]

Green, A. E. S., Green, B. A. S., Horvath, J. F., Samuels, J., Vaidya, D., Zeiler, B., and Pamidimukala, K. M., Field and Laboratory Experiments with Gas-Coal Flames (in progress, 1981). [47]

Hanson, H. P., Chapter 7, pp. 135-154, In Coal-Burning Issues, ICAAS, 1980. [126]

Hardesty, D. R., and Pohl, J. H., Sandia Laboratory Report 78-8804, 1979. [119]

Harris, E. R., Connell, G. F., and Dengiz, F., Coal Handling Equipment and Storage for Industrial Plants, Combustion, pp. 24-34, January 1976. [110]

Horton, M. D., Fast Pyrolysis, Chapter 8, Pulverized Coal Combustion and Gasification, Plenum Press, New York and London, 1979. [123]

Horvath, J. J., Vaidya, D. B., and Green, A. E. S., Observations of OH Emission (0,0 band) in a Methane-Coal Dust Flame, Optics News, 7, 3, p. 78, Oct. 26-30, 1981. [5]

Hottel, H. C. and Stewart, I. M., Space Requirement for the Combustion of Pulverized Coal, Ind. & Eng. Chem., 32, p. 719, 1940. [46]

ICAAS, Review of Gainesville/Alachua County. Regional Utilities Board's Application for Site Certification of Deerhaven Unit No. 2, prepared by Florida Department of Pollution Control by ICAAS, A. E. S. Green, P.I., Gainesville, Florida, February, 1975. [116]

ICAAS, An Interdisciplinary Study of the Health, Social and Environmental Economics of Sulfur Oxide Pollution in Florida, ICAAS Report, A. E. S. Green, P.I., February, 1978. [105]

ICAAS, Coal Burning Issues, Green, A. E. S. (Ed.), University Presses of Florida, Gainesville, Florida, January, 1980. [1, 19, 20, 23, 40, 115]

ICAAS, The Impact of Increased Coal Use in Florida (IICUF), Green, A. E. S. (Ed.), The University of Florida, Gainesville, Florida, November, 1980. [1, 20]

ICF, Inc., Boiler System Cost Estimates, Chapter 2, 1978. [14]

ICF, Inc., A Preliminary Analysis of Estimates of Emissions from Reconversion of Coal Capable Utility Boulers, prepared for Edison Electric Institute, April 23, 1980. [55, 101]

ICF, Inc., A Preliminary Economic Analysis of Reconversion of Coal Capable Utility Boilers, prepared for Edison Electric Institute, April 25, 1980. [68, 83, 85]

Jamgochian, E. M., Foo, O. K., and Skinner, T. F., Survey of Oil-Fired Utility Boilers, Potential for Coal-Oil Mixture Conversion, Mitre Corp. for U.S. Department of Energy, DOE/FE/53179-01, February, 1980. [108]

Kanury, A. M., Introduction to Combustion Phenomena, 2, Gordon and Breach, Science Publishers, Inc., September, 1975. [43, 45, 46, 47, 50]

Kellogg, W. W., Global Influence of Mankind on the Climate, Climate Change, J. Gibbon (Ed.), Cambridge University Press, 1978. [54]

Kellogg, W. W., and Schware, R., Climate Change and Society, Westview Press, 1981.[54]

Kidder, Peabody and Co., Fossil Power Status Reports, 1979. [130]

Kirkwood, J. B., Kregg, D. H., and Schoenwetter, H. D., Conversion to Coal of Gas/Oil-fired Boilers, Power Engineering, p. 66, July, 1978. [10, 52, 53, 65, 116]

Kirkwood, J. B., Commissioner's Corner, Energy Commission News, Western Australia, March, 1979. [10, 116]

Krickenberg, K. R., and Lubore, S. H., Coal Gasification: A Source of CO_2 for Enhanced Oil Recovery?, ES&T Features, Environmental Science & Technology, 15(12), pp. 1,418-1424, American Chemical Society, December, 1981. [39]

Kroptkim, P. N. and Valiaev, B. M., Development of a Theory of Deep-seated (inorganic and mixed) Origin of Hydrocarbon, in Goryuchie Iskopaemye, Problemy Geologii i Geokhimii Naftidov i Bituminoznykh Porod, pp. 133-144, Vassoevich, N. B. (Ed.), et al., Akademia Nauk SSSR, 1976. [37]

Landsberg, H. H. et al., Energy: The Next Twenty Years, Resources for the Future Study Group, Ford Foundation, Ballinger Publishing Co., Cambridge, Mass, 1979. [19, 115]

Lave, L. B. and Seskin, E. P., An Analysis of the Association Between U.S. Mortality and Air Pollution, Journal of American Statistics Association, 68, pp. 284-290, 1973. [105]

Little, J. W. and Capehart, L. C., Federal Regulatory and Legal Aspects, Chapter 18, pp. 359-381, In Coal Burning Issues, ICAAS, 1980. [131]

Loehman, E. F., Berg, S. V., Arroyo, A. A., Hedinger, R. A., Schwartz, J. M., Shaw, M. E., Fahien, R. W., De, V. H., Fishe, R. P., Rio, D. E., Rossley, W. F., and Green, A. E. S., Distributional Analysis of Regional Benefits and Cost of Air Quality Control, J. Environmental Economics and Management, 6, pp. 222-243, 1979. [105]

Maize, K., A Step Ahead for Burning Coal/Water Mix, The Energy Daily, 9, Oct. 15, 1981 [65]

Malte, P. C., and Rees, D. P., Mechanisms and Kinetics of Pollutant Formation During Reaction of Pulverized Coal, Chapter 11, Pulverized-Coal Combustion and Gasification, Plenum Press, New York and London, 1979. [123]

Mayster, S., Conversion of Industrial Plants to use Coal as Fuel, Mechanical Engineering, pp. 27-31, July, 1979. [110]

Measday, D. F. et al., Second International Conference on the Nucleon-Nucleon Interaction, Vancouver, British Columbia, American Institute of Physics, N.Y., 1978. [43]

Miller, H., and Modigliani, F., Some Estimates of the Cost of Capital to the Electric Utility Industry, 1954-57, in Modern Developments in Financial Management, pp. 152-210, Stewart C. Myers (Ed.), Hinsdale, The Dryden Press, 1976. [81]

Miller, D. J. and Lee, H. H., Synthetic Fuels From Coal, In Coal Burning Issues, Chapter 6, pp. 111-134, A. E. S. Green (Ed.), ICAAS, Jan., 1980. [39, 113]

Milner, M. R., Trimex Technology Report, Trimex Corporation, Clearwater, Florida, 1980. [104]

National Academy of Science, Energy and Climate, p. 158, Washington, D.C., 1977. [54]

National Academy of Science, Methane Generation from Human, Animal and Agricultural Wastes, Washington, D.C., 1977. [38]

National Academy of Science, Carbon Dioxide and Climate: A Scientific Assessment, p. 22, Washington, D.C., 1979. [54]

National Academy of Science, Energy in Transition 1985-2010, Final Report of the Committee on Nuclear and Alternative Energy Systems, W. H. Freeman and Co., San Francisco, California, 1980. [19, 115]

Nehring, R. and Van Priest, R., II, The Discovery of Significant Oil and Gas Fields in the United States, Rand Corporation, Santa Monica, California, 1981, (R-2654/1[/2])-USGS. [27]

Office of Technology Assessment, The Direct Use of Coal, Prospects, and Problems of Production and Combustion, Library of Congress, 79-600071, Washington, D.C., 1979. [115]

Ohanian, M. J., and Fardshisheh, F., Coal Availability and Coal Mining, Chapter 2, pp. 33-48. In Coal Burning Issues, ICAAS, 1980. [26]

Oppenheimer, E. J., Natural Gas: The New Energy Leader, Pen and Podium Productions, New York, N.Y., 1980. [37]

PEDCO Environmental, Inc., The Population and Characteristics of Industrial/Commercial Boilers, prepared for Environmental Protection Agency, May, 1979. [111]

Peters, W. and Schilling, H. D., An Appraisal of World Coal Resources and Their Future Availability, World Energy Resources, pp. 1485-2020. [23]

Philipp, J., and Kregg, D., Conversion from Oil to Coal-firing, Burns and Roe, Inc. Report, City of Gainesville, Department of Utilities, June 17, 1980. [10, 11, 53, 55, 62, 65, 73]

Philipp, J., Oil to Coal Conversion Options, Burns and Roe, Inc., Paper presented at EPRI/Florida Power and Light Seminar on Use of Coal in Oil-Design Utility Boilers, Lake Buena Vista, Florida, December 2-4, 1980. [10, 11, 53, 55, 62, 65, 67, 93]

Plank, D., Manager Engineering Department, City Utilities of Springfield, Missouri, telephone conversation, September, 1981. [47, 130]

Potential Gas Agency, Potential Gas Committee, Potential Supply of Natural Gas in the United States as of December 31, 1980, Golden Colorado, Colorado School of Mines, May, 1981. [29, 31]

Powerplant and Industrial Fuel Use Act of 1978, Public Law 95-620, 9th Congress, November 9, 1978. [86, 131]

Power Systems Services, Conversion to Pulverized Coal Firing, Tucson Electric Power Company, Irvington Station, Units 1, 2 & 3, November 17, 1980. [65]

Rawls, R., New Burner Reduces Nitrogen Oxide Emissions, Chemistry & Engineering News, p. 19, March 30, 1981. [104]

Rees, D. P., Lee, J., Brienza, A. R., and Heap, M. P., The Development of Distributed Mixing Pulverized Coal Burners, Proceedings of the Joint Symposium on Stationary Combustion NO_x Control, EPA/EPRI, 5, 249-273, October, 1980. [127]

Resources for the Future, A Note on Comparing Generating Costs. Energy in America's Future: Choice Before Us. pp. 272-273. The Johns Hopkins University Press, Baltimore, Maryland, 1980. [72, 81]

Rider, D. K., Energy: Hydrocarbon Fuels and Chemical Resources, John Wiley and Sons, New York, N.Y., 1981. [112]

Roach, C., Replacing Oil and Natural Gas with Coal: Prospects in the Manufacturing Industries, Congress of the United States, Congressional Budget Office, August, 1978. [110]

Rosenbaum, W. A., Coal and the States: A Public Choice Perspective, Chapter 17, pp. 343-358. In Coal Burning Issures, ICAAS, 1980. [131]

Schlenker, E. H., and Jaeger, M. J., Health Effects of Air Pollution Resulting from Coal Combustion, Chapter 14, pp. 277-302. In Coal Burning Issues, ICAAS, 1980.[131]

Schlesinger, B., Natural Gas Can Help Coal Burn Cleaner, American Chemical Society Journal, 14, pp. 1067-1071, 1980. [104]

Schmidt, R. A., Coal in America: An Encyclopedia of Reserves, Production and Use. McGraw-Hill, Inc., New York, N.Y., 1979. [25]

Shaw, S. H., An Industrial Boiler Fuel Modeling Approach, TR No. TR/IA/79-27, U.S. Energy Information Administration, 1979. [110]

Smith, P. L., Bagnall, L. O., and Nordstedt, R. A., Methane - Its Biological Production and Uses, Proceedings of Conference of Alternative Energy Sources for Florida, pp. 63-73, Gainesville, Florida, December 5-6, 1979. [38]

Smith, W. H., and Dowd, M. L., Journal of Forestry, 79, pp. 508-511, 1981. [38]

Smoot, L. D., and Pratt, D. T., Pulverized Coal Combustion and Gasification, Plenum Press, 1979. [45]

Smoot, L. D., Geoffrey, G. J., Cannon, J. N., Cutler, R. P., Schramm, D. E., Pulverized Coal Power Plant Fires and Explosions, Summary Report Part III, Chemical Engineering Department, Brigham Young University, October 15, 1980. [118]

Spaulding, D. B., Combustion of Fuel Particles, Fuel, 30, p. 121, 1951. [47]

Stout, H., Florida Gas Transmission Company, private communication, Aug. 18, 1981.[90]

Taylor, K. E., Chameides, W. L., and Green, A. E. S., Atmospheric Modifications, Chapter 11, pp. 203-230. In Coal Burning Issues, ICAAS, 1980. [131]

Tennessee Gas Transmission Company, Energy 1979-2000, Houston, Texas, May, 1980. [27]

Tennessee Gas Transmission Company, Energy 1981-2000, p. 18, Houston, Texas, August, 1981. [6]

Ten-Year Site Plans: Florida Power and Light, 1981; Jacksonville Electric, 1981; Orlando Utilities Commission, 1981; Gainesville Regional Utilities, 1981; City of Tallahassee, 1981; Florida Power Corporation, 1981; Tampa Electric Company, 1981; City of Lakeland, 1981; Gulf Power, 1981; Seminole Electric, 1981. [55, 89]

Thurgood, J., and Smoot, L. D., Volatiles Combustion, Chapter 10, in Pulverized-Coal Combustion and Gasification, L. D. Smoot and D. T. Pratt (Eds.), Plenum Press, New York, N.Y., 1979. [44]

Treadgold, J., World Lead in Oil-to-Coal Conversion, Engineers Australia, June 15, 1979. [116]

Ulrich, G. D., Milnes, B. A., and Subramanian, N. S., Particle Growth in Flames, II, Experimental Results for Silica Particles, Combustion Science Technology, 14, pp. 243-249, 1976. [124]

Urone, P. and Kenney, M. A., Atmospheric Pollution, Chapter 9, pp. 169-188. In Coal Burning Issues, ICAAS, 1980. [131]

U. S. Geological Survey, Estimates of Undiscovered Recoverable Resources of Conventionally Productible Oil and Gas in the United States, A Summary, Open-File Report 81-192,
Washington, D.C., 1981. [29]

Vunkov, A. E. et al., Combustion of Natural Gas in a Pulverized Coal Combustion Chamber,
Toploenergetika 3 (2), pp. 36-40, 1961. [47]

Wen, C. Y. and Dutta, S., Rates of Coal Pyrolysis and Gasification Reactions, Chapter
2, pp. 57-170, Coal Conversion Technology, C. Y. Wen and E. A. Lee (Eds.), Addison-
Wesley Publishing Co., Reading Mass., 1979. [123]

Wen, C. Y. and Lee, E. S., Coal Conversion Technology, Addison-Wesley, Reading, Mass.,
1979. [119, 121]

Whitehurst, D. D., 1978 American Chemical Society Symposium, 71, pp. 1-35, Organic
Chemistry of Coal, American Chemical Society, Washington, D.C., 1978. [50]

Wildavsky, A., and Tenenbaum, E., The Politics of Mistrust, Sage, Beverly Hills,
California, 1981. [131]

Wilkinson, P. L., American Gas Association Letter to A. E. S. Green, dated November 25,
1981. [6]

Wilson, S. U. et al., Florida Sulfur Oxide Study, Final Report of Florida Sulfur Oxide
Study, Post, Buckley, Schuh and Jernigan, Inc., Orlando, Florida, 1978. [100, 101]

Wingenroth, J., Telephone Conversation with Manager of Gas Supply, American Gas Association, October 26, 1981. [29]

Wingenroth, J., Xerox copy of table provided by Manager of Gas Supply, American Gas
Association, October 18, 1981. [29]

Wolfhard, H. G., Glassman, I., and Green, L., Jr. (Eds.), Heterogeneous Combustion:
Progress in Astronautics and Aeronautics, Academic Press, New York, 1964. [46]

Woltz, S. S., and Street, J. J., Agriculture, Chapter 13, pp. 249-276, In Coal Burning
Issues, ICAAS, 1980. [131]

Zaborsky, O. R. (Ed.-in-chief), CRC Handbook of Biosolar Resources, 2, Resource Materi-
al, National Science Foundation, Washington, D.C., T. A. McClure and E. S. Lipinsky
(Eds.), Battelle Columbus Laboratories, Columbus, Ohio, CRC Press, Inc., Boca Raton,
Florida, 1981. [18, 112]

Zinn, R. E., A Burner for Powders and Coal - Natural Gas Mixtures, Proceedings, Institute of Gas Technology, Bureau of Mines Symposium, Morgantown, W. Va., October
19-20, 1965. [47]

Subject Index

ERRATA

Table of Contents	Change "Foreword" to "Preface"
Page 8, Line 4	Change "700" to "800"
Line 5	Change "800" to "1000"
Page 11, Line 10	Change "third" to "fifth"
Line 12	Change "fourth" to "sixth"
Page 12, Table 1.4	Add "in million dollars" to caption
Page 15, Fig. 1.5	Change "50" on scale to "40"
Page 71, Table 5.4	Under "1980 Fuel Costs": Change Gas from "2.60/10^6CF" to "2.60/10^6 BTU"
Page 73, Line 8 ⎫ Page 76, Line 7 ⎬ Page 93, Line 10 ⎭	Change "million cubic feet" to read "million Btu"
Page 128, Table 11.1	Add to caption "We recommend"